ADVANCED NUMERICAL MODELS FOR SIMULATING TSUNAMI WAVES AND RUNUP

ADVANCES IN
COASTAL AND OCEAN ENGINEERING

ADVANCED NUMERICAL MODELS FOR SIMULATING TSUNAMI WAVES AND RUNUP

VOLUME 10

Editors

Philip L.-F. Liu
Cornell University, USA

Harry Yeh
Oregon State University, USA

Costas Synolakis
University of Southern California, USA

World Scientific

NEW JERSEY • LONDON • SINGAPORE • BEIJING • SHANGHAI • HONG KONG • TAIPEI • CHENNAI

Published by
World Scientific Publishing Co. Pte. Ltd.
5 Toh Tuck Link, Singapore 596224
USA office: 27 Warren Street, Suite 401-402, Hackensack, NJ 07601
UK office: 57 Shelton Street, Covent Garden, London WC2H 9HE

British Library Cataloguing-in-Publication Data
A catalogue record for this book is available from the British Library.

ADVANCED NUMERICAL MODELS FOR SIMULATING
TSUNAMI WAVES AND RUNUP

Copyright © 2008 by World Scientific Publishing Co. Pte. Ltd.

All rights reserved. This book, or parts thereof, may not be reproduced in any form or by any means, electronic or mechanical, including photocopying, recording or any information storage and retrieval system now known or to be invented, without written permission from the Publisher.

For photocopying of material in this volume, please pay a copying fee through the Copyright Clearance Center, Inc., 222 Rosewood Drive, Danvers, MA 01923, USA. In this case permission to photocopy is not required from the publisher.

ISBN-13 978-981-270-012-4
ISBN-10 981-270-012-9

Printed in Singapore.

Preface

This review volume grew out of THE THIRD INTERNATIONAL WORKSHOP ON LONG-WAVE RUNUP MODELS that took place June 11–12, 2004 at the Wrigley Marine Science Center of the University of Southern California on Catalina Island, California. This workshop and the two previous workshops were sponsored by the Geo-Hazard program of the National Science Foundation of the United States. Dr. Clifford J. Astill was the director of the program.

The first long-wave runup workshop was also held in the Wrigley Marine Science Center, on August 15–17, 1990. Tsunami runup/inundation and the hazard mitigation were the main topics discussed in the first workshop, which brought together researchers from Japan, Russia, England, and the United States. The proceedings of the workshop were reported by Liu, et al. (1991). The workshop revitalized research efforts on long-wave runup, spawning several major research programs on long-wave runup in the United States as well as overseas. The workshop has also facilitated further international research collaborations.

From 1992 to 1994 several major tsunamis occurred, including the Nicaraguan tsunami in 1992; the Flores Island (Indonesia) tsunami in 1992; the Okushiri (Japan) tsunami in 1993. All three tsunamis caused devastating property damages and many deaths. Moreover, in 1994 alone four additional tsunamis, including the East Java (Indonesia), the Shikotan Island (Russia/Japan), the Mindoro (Philippines), and the Skagway (Alaska, USA) tsunamis, occurred around the world. The runup heights along affected coastlines were surveyed and documented by various research teams. Different research groups have performed numerical simulations of all these events with different numerical models. The community felt the strong need to gather researchers together to discuss similarities and dissimilarities among these models and to discuss the issues in modeling coastal effects of tsunami.

Consequently, the Second International Workshop on Long-Wave Runup Models was held at Friday Harbor, San Juan Island, Washington, on September 12–16, 1995. The participants in the workshop were from Japan, Russia, England, Brazil, Australia, Canada, Indonesia, and the United

States. The format of the second workshop was designed to focus more on discussions than on formal presentations. Four benchmark problems were selected before the workshop so that numerical models can be compared, evaluated and discussed among the participants. During the workshop, seven discussion themes were organized as follows: laboratory, analytical, finite-difference, finite-element, vertical-plane models, boundary-integral-element models, and marker-and-cell models. The presentations and discussions were edited and published in a book entitled "Long-Wave Runup Models" (Yeh, et al. 1996).

There was no doubt that the benchmark-problem exercises used in the second workshop proved extremely useful in identifying absolute and comparative modeling capabilities. Overall, in terms of tsunami runup modeling, significant advances had been made between two workshops, due mainly to the advancement of computational capabilities, and because of the generation of a large 2-D and 3-D laboratory data set and the fortuitous field measurements in 1992–1995, all of which have contributed to model calibrations. The tsunami modeling efforts had become more directed towards their implementation for real tsunami predictions and hind-castings than ever before. At the time of the 1990 Catalina workshop, large differences between computed runup results and field measurements might have been attributed to both errors in the seismic estimates of the source motion and to the hydrodynamic calculations. Much advances had been made by the 1995 Friday Harbor workshop; researchers became much more confident in the hydrodynamic calculations, at least for non-breaking waves. It was equally clear that reduction and even elimination of numerical dispersion and numerical dissipation effects would — if not already — soon become reality. At the same time, the workshop participants recognized additional and important problems arising from modeling improvements, such as determination of highly accurate initial wave conditions, modeling the three-dimensional flow effects, and turbulence. Actual tsunami runup motions are turbulent; the runup flow patterns, impacts, scouring effects, and sediment transport are all affected by turbulence in the runup motions.

From 1995 to the summer of 2004, there have been at least six additional large tsunamis resulting in catastrophic loss of life and property. They are the Irian Jaya tsunami (Indonesia) in 1996, the Peru tsunamis in 1996 and in 2001, the Papua New Guinea tsunami in 1998, the Turkey tsunami in 2000, and the Stromboli (Italy) tsunami in 2002. Among these six tsunamis, the Turkey tsunami was definitely caused by land subsidence and slides associated with earthquakes, and the Stromboli tsunami was caused by landslides caused by volcanic eruption. On the other hand, the source of the Papua New Guinea tsunami, which killed more than 2000 people and destroyed completely three villages, remains controversial and has been postulated as due either to co-seismic seafloor dislocation or sediment slump. Because of the occurrence of these tsunamis, research interest and efforts on the modeling of landslide generated tsunamis have been intensified in recent years. The landslide-generated tsunamis have very different characteristics from the tsunamis generated by earthquakes. The traditional depth-integrated shallow water equations are not always adequate for modeling the landslide generated tsunamis. Other noteworthy advances in recent years are the development of several computational fluid dynamics models calculating the nearshore waves and their interactions with structures with the consideration of frequency dispersion and turbulence.

The primary objectives of the third workshop were to provide a platform for discussing both old and new numerical models and their applications to various critical issues concerning tsunami runup and wave-structure interactions. To accomplish this goal, four benchmark problems were selected and posted on the workshop website before the workshop. http://www.cee.cornell.edu/longwave/. The workshop participants were given the solutions, in the form of laboratory data or analytical solutions, a week before the workshop. The workshop participants discussed their numerical model and the comparisons between their numerical results and solutions for the benchmark problems. Their presentations are also posted on the workshop website. The four benchmark problems are as follows:

1. Calculations of the moving shoreline.
2. Tsunami runup onto a complex three-dimensional topography.
3. Landslide generated tsunami.
4. Tsunami forces on a nearshore structure.

This review volume is divided into two parts. The first part includes five review papers on various numerical models. Pedersen provided a brief but thorough review on the theoretical background for depth-integrated wave equations, which are employed to simulate tsunami runup. LeVeque and George describe high-resolution finite volume methods for solving the nonlinear shallow water equations. They have focused their discussion on the applications of these methods to tsunami runup. In recent years, several advanced 3D numerical models have been introduced to the field of Coastal Engineering to calculate breaking waves and wave-structure interactions. These models are still being developed and are at different stage of maturity. Roger and Dalrymple discussed the Smooth Particles Hydrodynamics (SPH) method, which is a meshless method. Wu and Liu presented their Large Eddy Simulation (LES) model for simulating the landslide generated waves. Frandsen introduced the Lattice Boltzmann method with the consideration of a free surface.

The second part of the review volume contains the descriptions of the benchmark problems and twelve extended abstracts submitted by the workshop participants. All these papers compared their numerical results with benchmark solutions.

Professor Synolakis and his staff of the University of Southern California were responsible for the local arrangements. Their efforts are deeply appreciated. The workshop was sponsored by the US National Foundation.

Philip L.-F. Liu, Harry Yeh and Costas Synolakis

References

Liu, P. L.-F., Synolakis, C. and Yeh, H., A report on the international workshop on long wave runup, *J. Fluid Mech.* **229** (1991), 678–88.

Yeh, H., Liu, P. L.-F. and Synolakis, C. (ed.), *Long-Wave Runup Models* (World Scientific, 1996), 403p.

A Dedication to Dr. Cliff J. Astill

This workshop was the third of a series of international workshops, which discussed the state of arts of analytical and numerical models for tsunami propagation and runup. The first workshop was held in 1990 on Catalina Island — the same venue as the present workshop — followed by the second workshop in 1995 at Friday Harbor, Washington. All three workshops have been supported by the Geo-Hazard program of the Civil and Mechanical Systems Division at the US National Science Foundation, which was under the directorship of the late Dr. Cliff J. Astill, who passed away on 23rd of December 2004 after a long and courageous fight against cancer. During his tenure as the Program Director, Dr. Astill was a proactive supporter of tsunami research, which has not gained as much attention as other earthquake research areas do. Coincidentally two days after Dr. Astill's death, the Great Indian Ocean Tsunami tragically struck the entire Indian Ocean, sending tsunami awareness around the globe and urging more accelerated research in tsunamis. This book is dedicated to Dr. Astill with our deep appreciation to his effort in promoting and guiding tsunami research.

Dr. Astill was born on March 4, 1936 in Brisbane, Australia. He received his Bachelor of Engineering degree in Mechanical Engineering from the University of Queensland in 1957. After the graduation, he worked in both weapons research establishments and aircraft design industry for almost 11 years. In 1965 Dr. Astill immigrated to the United States after working in England and Canada. He received his Master of Engineering and Doctoral of Science (ScD) degrees in Engineering Mechanics from Columbia University in 1968 and 1971, respectively.

Dr. Astill started working at NSF in June 1971 as an associate program director in the Engineering Mechanics Program, which was then under the Research Directorate. He became the Program Director for Solid Mechanics Program in the Engineering Division in the late seventies and then the Program Director of Sitting & Geotechnical Systems, Earthquake Hazard Mitigation in the Division of Civil and Mechanical Systems in the early eighties. The program has gone through several name changes in the eighties and nineties. Although NSF started funding tsunami research after the 1960 Chilean tsunami, tsunami research had not received consistent and steady funding until Dr. Astill took over the program. Dr. Astill's enthusiasm and steady support in tsunami research over the last 20 years is the catalyst for all the achievement made to date. He has indeed been a mentor for the US tsunami research community until his retirement from NSF in January 2004.

Dr. Astill had a vision that tsunami research ought to be interdisciplinary. On one hand, he supported the fundamental research in developing mathematical models and performing basic laboratory experiments and numerical simulations for various aspects of physical processes related to tsunami generation, propagation and runup. On the other hand, he recognized the importance of the hazard mitigation practice by supporting research in the areas of social sciences, policy-making and planning, and data management. Fifteen years ago, he initiated such multi-disciplinary activities by strongly encouraging researchers to submit cooperative group proposals, which were unheard of in those days. Dr. Astill was always proactive in promoting promising young researchers. In many occasions, using himself as a source of information, he introduced young researchers to senior researchers and encouraged them to form a possible collaborative team.

Dr. Astill also firmly believed that an effective means to develop multi-discipline and cross-disciplinary research programs is through informal small group meetings and formal workshops. Over the years he has supported a numerous such meetings and workshops (see the list below). Through his encouragement, the tsunami community has expanded its membership to include not only hydrodynamists and seismologists but also geologists, social scientists, oceanographers, mathematicians, and data managers. When we look back, Dr. Astill led us to the research mode of "collaboratory" more than 15 years ago, which is the fundamental philosophy of NSF NEES project (Network for Earthquake Engineering Simulations: http://www.nees.org/index.php). Dr. Astill was not afraid of offering his opinions and making suggestions to workshop organizers. For example, the procedure of "Blind Tests" in validating numerical models was first recommended by Dr. Astill and implemented in the 1995 Second International Workshop on Long-Wave Runup Models at Friday Harbor. Prior to the workshop — April 1994 — Dr. Astill wrote:

> "I don't know the details, so the following remarks may not be relevant. However in a recent effort in the area of using blind predictions (Class A experiments) to validate the procedures used to determine the liquefaction potential of soils, there

were some surprises. (Cliff was referring to VELACS project (Verification of Liquefaction Analysis by Centrifuge Studies): http://www.cee.princeton.edu/ radu/soil/velacs/) First, I should explain these Classes of predictions. Class A: A "single-blind" experiment — using the parlance from the world of medical research — where the predictions are made before the experiments are run. Class B: Predictions made AFTER the experiment is run, but with the experimental results not known to the predictor. Class C: The predictor knows the results of the experiments to be "predicted". There is no class, as far as I know, corresponding to the "double-blind" experiment of the medical researcher. There is a role for each of these classes. In the exercise referred to, after the experiment was run, the predictors were given the actual configuration of the experiment (the prediction was based on the nominal configuration), but these results were then Class B predictions: the experiment had been run, but the predictors were still unaware of the results from the experiment. Finally, there was still a lot of work to do after the experimental results were made public, and these were then classified as Class C predictions. Well, so much for naive expectations. In practice, there were accusations that some predictors had used the results from prior similar experiments; even that some, who were also experimentalists, had actually run the experiment themselves according to the nominal configuration on which they were to make their predictions! As I said, maybe none of this is germane to your proposed endeavors. But if it is, do you have specific plans to try to guarantee that blind predictions are truly blind?

This reminds me of an incident in my early days at NSF, which overlapped with the last year or two of the mini-skirt fashions. A blind flower seller used to sit just outside our building on G Street, selling flowers. One day, a young lady in an extremely short dress was observed to walk by this blind flower seller, whose head tracked her passing like it was locked on by a radar beam. Subsequently, I discovered that there are different degrees of being officially blind. To be classified as officially blind does not imply the total absence of sight. So there is the warning."

Dr. Astill was a strong supporter for the reconnaissance field surveys for tsunamis. Since tsunamis are rare and high impact events, lessons learned from the field surveys are invaluable. Dr. Astill believed that the strong links between the real-world observations and the laboratory and/or theoretical understandings are crucial for natural hazards research. Based on the field observations, many topics of fundamental research were identified and several theoretical hypotheses were validated. Dr. Astill's effort to support reconnaissance tsunami survey began after the 1992 Nicaragua Tsunami, which was the first major tsunami since 1983 Nihonkai Chubu Tsunami. He developed several options, which enabled researchers to quickly go to the filed without being burden by paper works. He clearly understood the importance of the survey and its critical timing for capturing perishable data. The tsunami community made a numerous tsunami surveys during his tenure (see the list below), which resulted in substantial advances in tsunami research.

Dr. Astill also proactively encouraged international collaboration by inviting international experts to serve on the advisory committees for several group research projects. He also suggested that we should participate more in international conferences and meetings: for example, the International Travel Grants to attend the General Assembly of European Geophysical Society in 2001 and 2002 were funded by Dr. Astill.

Dr. Astill's career at NSF can be best described by his own words on May 31, 2004.

"Growing up in Australia was a miserable experience, in retrospect. My life changed unexpectedly once I left Australia and although my life has been very prosaic, those 44 years have been like a fairy tale compared to the 24 years spent in Australia, and that especially includes the 32 years at NSF. I took so much for granted while I was at NSF, except for that last year, when it became painfully obvious, based on the hiring in CMS during that year, that I would not have been hired if I had applied for my own position. At first that was a real shock, but then I began to realize just how lucky I had been to be able to spend 32 years there; 14 years running the Solid Mechanics Program, and 18 years running the geotechnical earthquake program. While I very much enjoyed those 14 years with the Solid Mechanics program, and especially the rigor of the continuum mechanic expertise that the researchers possessed, and was sorry to leave it, I nonetheless found the earthquake researchers to be a much nicer group of people to interact with. I very much missed the expertise in continuum mechanics in geotechnical engineering, and was most grateful to the tsunami community for having that expertise. How come that civil engineers specializing in soil mechanics receive such an abysmal training in continuum mechanics, while those specializing in fluid mechanics receive it? .... I've been very surprised at the way my thoughts on NSF basically evaporated the day I retired. After spending 32 years there, this seems almost unreal, but it has happened. I've given a lot of thought to it, and the fact is that we all sooner or later lose touch and are forgotten. ...... All in all, the outlook looks very gloomy except for NEES."

Dr. Astill was a visionary in tsunami research. He had guided and supported the tsunami research community in the last twenty years. We were very fortunate to have the privilege to know him professionally and personally. His open-minded way in dealing with professional issues and his calm spirits in fighting cancer will forever stay with us. He will be terribly missed by all of us.

Harry Yeh, Philip Liu and Costas Synolakis

List of Workshops Supported by Cliff:

Hydrodynamic Model Developments:

- o International Workshop on Long-Wave Run-up, Catalina Island, CA, August 1990
- o Second International Workshop on Long-Wave Runup Models, Friday Harbor, San Juan Island, WA, September 1995
- o The Third International Workshop on Long-Wave Runup Models, Catalina Island, CA, June 2004

Development of Hazards Mitigation Tools:

- o Workshop on the Development of a Tsunami Scenario Simulation Program, Seattle, WA, September 2002
- o Workshop for an Integrated Tsunami Scenario Simulation, Corvallis, OR, August 2003 & San Francisco, CA, October 2004

Tsunami Field Survey and Data Management:

- o Tsunami Measurement Workshop, Estes Park, CO, June 1995
- o International Workshop on Bathymetry and Coastal Topography Data Management, Seattle, WA, March 1998 & Birmingham, UK, July 1999

Tsunami Sources:

- o Workshop on Seafloor Deformation Models, Santa Monica, CA, May 1997
- o Workshop on the Prediction of Underwater Landslide and Slump Occurrence and Tsunami Hazards, Los Angeles, CA, March 2000

Development of NEES:

- o Workshop for Tsunami Research Facilities, Baltimore, May 1998
- o Ad-Hoc Planning Working Group for the NSF Major Research Equipment Program: National Earthquake Engineering Simulations, 1997–1999
- o Workshop on Research with NEES Tsunami Facility, Corvallis, OR, March 2001

List of Tsunami Surveys Supported by Cliff:

Nicaragua Earthquake and Tsunami, September 1992
Flores-Island (Indonesia) Earthquake and Tsunami, December 1992
Okushiri-Island (Japan) Tsunami, July 1993
The East Java Tsunami (Indonesia), June 1994
South-Kuril-Island (Russia/Japan) Earthquake and Tsunami, October 1994
Mindoro (Philippines) Tsunami, November 1994
Skagway (Alaska) Landslide Tsunami, November 1994
Chimbite (Peru) Tsunami, February 1996
Irian Jaya (Indonesia) Tsunami, February 1996
Aitape (Papua New Guinea) Tsunami, July 1998
Vanuatu Tsunami, November 1999
Southern Peru Tsunami, June 2001

List of International Travel Grants Supported by Cliff:

HAZARD-2000, Tokushima, Japan, May 2000
HAZARD-2002, Antalya, Turkey, September 2002
Tsunami Workshop in Petropavlovsk-Kamchatskiy, Kamchatka, Russia, in August 1996
European Geophysical Society — XXVII General Assembly, Nice, France, April 2001 & March 2002

A Dedication to Dr. Cliff J. Astill

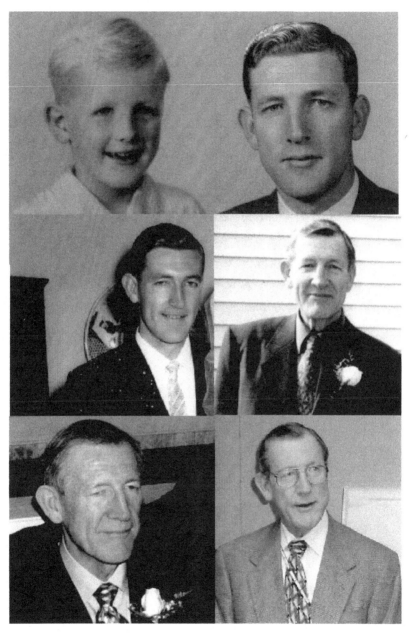

Clifford John Astill
March 4, 1936 – December 23, 2004

Contents

Preface ... v

A Dedication to Dr. Cliff J. Astill ... ix

Part 1: Review Papers

Chapter 1
Modeling Runup with Depth Integrated Equation Models ... 3
  G. Pedersen

Chapter 2
High-Resolution Finite Volume Methods for the Shallow Water
Equations with Bathymetry and Dry States ... 43
  R. J. LeVeque and D. L. George

Chapter 3
SPH Modeling of Tsunami Waves ... 75
  B. D. Rogers and R. A. Dalrymple

Chapter 4
A Large Eddy Simulation Model for Tsunami and Runup
Generated by Landslides ... 101
  T.-R. Wu and P. L.-F. Liu

Chapter 5
Free-Surface Lattice Boltzmann Modeling in Single Phase Flows ... 163
  J. B. Frandsen

Part 2: Extended Abstracts

Chapter 6
Benchmark Problems ... 223
  P. L.-F. Liu, H. Yeh and C. E. Synolakis

Chapter 7
Tsunami Runup onto a Plane Beach 231
    Z. Kowalik, J. Horrillo and E. Kornkven

Chapter 8
Nonlinear Evolution of Long Waves over a Sloping Beach 237
    U. Kânoğlu

Chapter 9
Amplitude Evolution and Runup of Long Waves;
Comparison of Experimental and Numerical Data
on a 3D Complex Topography 243
    A. C. Yalciner, F. Imamura and C. E. Synolakis

Chapter 10
Numerical Simulations of Tsunami Runup onto a
Three-Dimensional Beach with Shallow Water Equations 249
    X. Wang, P. L.-F. Liu and A. Orfila

Chapter 11
3D Numerical Simulation of Tsunami Runup onto a
Complex Beach 255
    T. Kakinuma

Chapter 12
Evaluating Wave Propagation and Inundation Characteristics
of the Most Tsunami Model over a Complex 3D Beach 261
    A. Chawla, J. Borrero and V. Titov

Chapter 13
Tsunami Generation and Runup Due to a 2D Landslide 269
    Z. Kowalik, J. Horrillo and E. Kornkven

Chapter 14
Boussinesq Modeling of Landslide-Generated Waves and Tsunami
Runup 273
    O. Nwogu

Chapter 15

Numerical Simulation of Tsunami Runup onto a Complex Beach
with a Boundary-Fitting Cell System 279
   H. Yasuda

Chapter 16

A 1-D Lattice Boltzmann Model Applied to Tsunami Runup onto a
Plane Beach 283
   J. B. Frandsen

Chapter 17

A Lagrangian Model Applied to Runup Problems 311
   G. Pedersen

Appendix

Phase-Averaged Towed PIV Measurements for Regular Head Waves
in a Model Ship Towing Tank 319
   J. Longo, J. Shao, M. Irvine, L. Gui and F. Stern

Part 1
Review Papers

CHAPTER 1

MODELING RUNUP WITH DEPTH INTEGRATED EQUATION MODELS

G. Pedersen

Mechanics Division, Department of Mathematics, University of Oslo
E-mail: geirkp@math.uio.no

This survey starts with a sketch of long wave theory, reflecting some of its diversity. Next, general aspects of runup on sloping beaches are briefly discussed in the scope of long waves, including important analytical solutions, breaking criteria, significance of non-hydrostatic effects and experiments. Then the literature on numerical modeling of runup is reviewed. The models are loosely organized into classes and then described chronologically within each class. Primarily, we refer models that are published in international journals, while matter from the vast number of proceedings and internal reports are mentioned occasionally. Moreover, we highlight the verification of the numerical methods by comparison to analytical solutions or experiments. Applications to real tsunami or storm surge events are, with a few exceptions, not included.

1. Introduction

Comprehension of wave runup on beaches is essential for prediction of beach erosion and coastal impact of tsunamis and storm surges. This is one reason for the lasting attention that runup topics have received in the literature of hydrodynamics and coastal engineering. Moreover, this complicated, yet commonplace phenomenon, that can be observed at every beach with incident swells or wind waves, pose mathematical and conceptual challenges that appeal to many scientists. Wave runup also inherits similarities to other important phenomena in hydrodynamics, such as "green water" (wave overtopping on vessels), slamming and interaction between the ocean pycnocline and bathymetry.

As for many other wave phenomena, long wave theory has been crucial for our understanding of wave runup on beaches and derivation of closed form solutions supporting this understanding. Also numerical models for large scale ocean waves in general, such as tides, storm surges and tsunamis, have traditionally been based on shallow water theory, due to the simplicity and efficiency of this formulation. Certainly, more general models are in progress, also for runup on beaches[23–25,43,50,51,70,109], but do still involve heavy computations. For instance, most fair sized three-dimensional wave problems are still dependent on approximate methods, such as long wave theory. In fact, even an idealized, two-dimensional simulation modeling the propagation of a tsunami wave or swell from deep water, through shoaling/surf and to runup on a shore will often border on the limits of models based on the Navier-Stokes equations or full potential theory. Therefore, for many applications we must rely on long wave models also in the years to come.

In the context of depth integrated equations, the moving shoreline on a beach is in some respects similar to a free water surface that is deformed by waves; the flow field is confined by a time dependent boundary with a position that is unknown *a priori*. Furthermore, in both cases the surface excursions correspond to motion of material particles with gravity as restoring force. However, there are important differences. In runup the medium is "thinning" toward the shoreline, resulting in a singularity in the governing equations, though not in the physically relevant solutions. In addition, wave breaking and topographical effects are often important for the near-shore flow. If we look beyond the long wave (depth-integrated) description, other aspects appear, such as the dynamics of a contact point between a free surface and a rigid no-slip boundary, boundary layers, and effects of bottom porosity. These are presumably most important on laboratory scale and are anyway outside the scope of the present article.

An early review on long wave runup is found in Meyer and Taylor (1972)[67]. The predecessors to the "The Third International Workshop On Long-Wave Runup Models", leading to this volume, also initiated articles summarizing aspects of runup research[53,113]. Among the contributions are overviews of finite difference and element methods for tsunamis, that also address long wave runup[32,105].

This review is meant to provide a brief and updated status on long wave runup models, with an inclination toward applicability of the different long wave approximations. We do not include related themes such as runup on vertical walls, edge waves, rip currents, and impact problems. Moreover, we focus on the principal aspects of physics and modeling rather than case studies of real events of tsunamis and storm surges. Wave breaking is the subject of Chapter 2 (by LeVeque and George) in this volume[48]. Still, for completeness, models with breaking, as well as theory and experiments, will be discussed briefly herein.

We start with a short presentation of relevant parts of long wave theory. Then the runup phenomenon is discussed with selected analytic solutions and experiments, before the diversity of models in the literature will be summarized. No complete assessment of the various methods will be attempted, but documentation of performance through comparison with analytic solutions, experiments, other models and grid refinement will generally be emphasized. Significance of physical processes like wave dispersion will also be addressed when appropriate.

2. Long Wave Equations

Following standard conventions we employ a typical depth d and a wavelength L as vertical and horizontal length scales, respectively. The choice of L, in particular, is ambiguous and it may also correspond to other lengths than that of a wave. Identifying the time scale $t_c = L/\sqrt{gd}$, where g is the acceleration of gravity, and the dimensionless amplitude, ϵ, we define $\epsilon L/t_c$, $\epsilon d/t_c$ and ϵd, respectively, as scales for horizontal velocity, vertical velocity and surface elevation. When we instead refer to dimensional quantities they are marked by a star. Different long wave equations can be obtained through perturbation expansions in $\mu \equiv d/L$ and ϵ and may then be classified according to which orders of these parameters that are retained in the equations. Omission of all μ terms yields the nonlinear shallow water (NLSW) equations, while retaining second-order in μ yields Boussinesq type equations that are available in a series of varieties[61,79,110]. Long wave approximations prescribe simple vertical structures for the field variables. In NLSW theory, the horizontal velocity is vertically uniform and the pressure hydrostatic. Therefore, the vertical coordinate, z, vanishes from the equations. When $O(\mu^2)$ terms are retained there are vertical variations in the horizontal velocity. Still, the explicit appearance of z is removed from the continuity and momentum equations by integration. Hence, we denote them as "depth integrated equations". The spatial dimension of the partial differential equations is then reduced by one. Furthermore, the nonlinear free surface appears only through nonlinear coefficients. These features make long wave formulations well suited for numerical solution. Regardless of the mathematical reduction of dimension, we will still refer to problems as two- or three-dimensional according to the physical configuration.

The order of a long wave equation is reflected in the dispersion characteristics. According to full potential theory the dispersion relation of a linear, sinusoidal wave reads (depth: $h=1$)

$$c^2 = \frac{1}{k\mu}\tanh(\mu k) = 1 - \frac{1}{3}(\mu k)^2 + \frac{2}{15}(\mu k)^4 + ..., \qquad (1)$$

where c is the phase speed and k is the wave number. Shallow water theory only reproduces the first term on the right hand, while traditional Boussinesq equations yield the first two. However, as we will see below, different formulations valid to $O(\mu^2)$ give different dispersion properties. They may even reproduce the $O(\mu^4)$ term in (1) correctly or display en extended validity range.

2.1. *Boussinesq Theory*

During the last fifteen years fully nonlinear Boussinesq type equations with improved dispersion properties have been put to work in computer models. Following Hsiao *et al.* (2002)[29] we write a set of fully nonlinear Boussinesq equations on the form

$$\eta_t = -\nabla \cdot \left[(h+\epsilon\eta)(\mathbf{v}+\mu^2\mathbf{M})\right] + O(\mu^4), \qquad (2)$$

$$\mathbf{v}_t + \frac{\epsilon}{2}\nabla(\mathbf{v}^2) = -\nabla\eta - \mu^2\left[\frac{1}{2}z_\alpha^2\nabla\nabla\cdot\mathbf{v}_t + z_\alpha\nabla\nabla\cdot(h\mathbf{v}_t)\right]$$
$$+ \epsilon\mu^2\nabla(D_1 + \epsilon D_2 + \epsilon^2 D_3) + O(\mu^4) + \mathbf{N} + \mathbf{E}, \qquad (3)$$

with

$$\mathbf{M} = \left[\frac{1}{2}z_\alpha^2 - \frac{1}{6}(h^2 - \epsilon h\eta + \epsilon^2\eta^2)\right]\nabla\nabla\cdot\mathbf{v} + \left[z_\alpha + \frac{1}{2}(h-\epsilon\eta)\right]\nabla\nabla\cdot(h\mathbf{v}),$$

$$D_1 = \eta\nabla\cdot(h\mathbf{v}_t) - \frac{1}{2}z_\alpha^2\mathbf{v}\cdot\nabla\nabla\mathbf{v} - z_\alpha\mathbf{v}\cdot\nabla\nabla\cdot(h\mathbf{v}) - \frac{1}{2}(\nabla\cdot(h\mathbf{v}))^2,$$

$$D_2 = \frac{1}{2}\eta^2\nabla\cdot\mathbf{v}_t + \eta\mathbf{v}\nabla\nabla\cdot(h\mathbf{v}) - \eta\nabla\cdot(h\mathbf{v})\nabla\cdot\mathbf{v},$$

$$D_3 = \frac{1}{2}\eta^2\left[\mathbf{v}\cdot\nabla\nabla\cdot\mathbf{v} - (\nabla\cdot\mathbf{v})^2\right],$$

and where the index t denotes temporal differentiation, h is the equilibrium depth, η is the surface elevation, $\mathbf{v}$ is the horizontal velocity evaluated at $z=z_\alpha(x,y)$ and ∇ is the horizontal component of the gradient operator. The heuristic terms $\mathbf{N}$ and $\mathbf{E}$ in the momentum equation (3) represent bottom drag and artificial diffusion and will be discussed below. This particular form of the leading dispersion ($\epsilon^0\mu^2$) term was discussed and tested by Nwogu (1993)[71], while additional nonlinearity was added by Wei *et al.*

(1995)[107]. A similar formulation, with the velocity potential as primary unknown instead of the velocity, is found in Chen and Liu (1995)[7]. Applying transformations and deletion of various higher order terms to the set (2) and (3), we may reproduce a number of other Boussinesq type equations from the literature[42,71,107]. As pointed out in the references[107] weakly nonlinear versions, in the sense that some or all $O(\epsilon\mu^2)$ terms are omitted, may sometimes yield nonzero volume flux at the shore. This should in particular be avoided for runup simulations.

When h is constant and $\epsilon \to 0$ Boussinesq equations on the form (2,3) yield c^2 as a rational function of $\mu^2 k^2$. For $z_\alpha = (\sqrt{1/5}-1)h$ this expression reproduces the first three terms in the expansion (1), while $z_\alpha = -0.531h$ yields a particularly favorable dispersion relation over an extended range of wave numbers. (see Fig. 1). With $z_\alpha = (\sqrt{1/3} - 1)h$, the velocity **v** differs from the depth averaged velocity ($\overline{\mathbf{v}}$) by $O(\epsilon\mu^2)$, only. If we then delete all $O(\epsilon\mu^2)$ terms in (2) and (3) we retrieve the traditional Boussinesq equations for constant depth. These possess a dispersion relation that is clearly inferior to the optimal version of (2,3) (Fig. 1). In general, the form of the volume flux in (2) implies[a] $\overline{\mathbf{v}} = \mathbf{v} + \mu^2 \mathbf{M}$. Differentiating this relation with respect to time, inserting the resulting expression for $\mathbf{v}_t$ in (3), and invoking $\mathbf{v} = \overline{\mathbf{v}} + O(\mu^2)$ in nonlinear and dispersive terms we obtain the standard Boussinesq equations that inherit errors of order $\epsilon\mu^2, \mu^4$

$$\eta_t = -\nabla \cdot [(h + \epsilon\eta)\overline{\mathbf{v}}], \tag{4}$$

$$\overline{\mathbf{v}}_t + \frac{\epsilon}{2}\nabla(\overline{\mathbf{v}}^2) = -\nabla\eta + \mu^2 \left[\frac{h}{2}\nabla\nabla \cdot (h\overline{\mathbf{v}}_t) - \frac{h^2}{6}\nabla\nabla \cdot \overline{\mathbf{v}}_t\right]$$
$$+ O(\epsilon\mu^2, \mu^4) + \mathbf{N} + \mathbf{E}. \tag{5}$$

This set has been used much in numerical simulations and for the first time by Peregrine (1967)[78] for the shoaling of a solitary wave. We will subsequently refer to formulations like (4,5) as "standard" Boussinesq equations, whereas those with more accurate dispersion properties, such as (2,3) with $z_\alpha = -0.531h$, will be denoted as "improved".

Further information on, and extensions of, higher order long wave equations are found in a number of papers in the literature[16,58-61]. However, progress has primarily been made with respect to higher order nonlinearity

---

[a]To reproduce $\overline{\mathbf{v}}$ from **v** in non-constant depth we would have to employ a time dependent z_α.

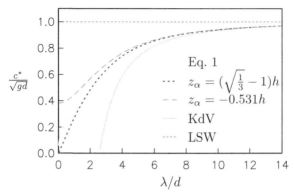

Fig. 1. The phase speed as function of wavelength for long wave equations compared to that of the fully inviscid set. The curve for the Korteweg-deVries (KdV) equation is included for comparison.

and dispersion during propagation in mildly varying bathymetries. Corresponding improvement has not yet been reported for a rapidly varying bottom. The shoreline, with its mathematical singularity and abrupt change of wave characteristics, as discussed below Eq. (7), may hardly be regarded as belonging to a region of mild slope. Hence, while the higher order Boussinesq formulations are of great value in finite depths, it remains to be seen what may be locally achieved in runup on a sloping beach. In fact, as will be discussed subsequently, even the significance of leading order dispersion is debatable in the vicinity of the shoreline.

2.2. Shallow Water Equations

Deletion of all $O(\mu^2)$ terms simplifies (2,3) to the NLSW equations. The pressure gradient in the momentum equation then becomes $\nabla \eta$, corresponding to hydrostatic pressure. As a consequence the horizontal velocity becomes independent of z. The NLSW equations lack the important effect of wave dispersion and may lead to erroneous results for long term propagation, even if the waves are long compared to the depth[79]. However, they are still quite a reasonable option for surf zone dynamics and are, by far, the most commonly used framework for runup calculations, as well as tsunami and storm surge models. Also the NLSW equations may be reformulated in different ways. For modeling shocks (bores), in particular, it is important to recast the momentum equations into a conservative form. When F and B are omitted we may write the x-component as

$$(Hu)_t + (\epsilon H u^2 + \frac{1}{2}H^2)_x + \epsilon(Huv)_y = Hh_x, \qquad (6)$$

where subscripts indicate partial derivation, $H = h + \epsilon\eta$, and u and v are the velocity components in the x- and y-directions, respectively. Naturally, a corresponding equation applies to the y-direction. We observe that the right hand side is a momentum source due to a sloping bottom. Often the depth integrated fluxes are used as primary unknowns instead of the velocities. This may cause difficulties during runup, since velocities are needed for the propagation of the shoreline and the relation between fluxes and velocities may be poorly defined when $H \to 0$.

A popular transformation of the NLSW equations is obtained through introduction of characteristics. For plane waves the characteristic equations become

$$\beta_t^\pm + (\epsilon u \pm c)\beta_x^\pm = h_x, \quad \text{where} \quad \beta^\pm = \epsilon u \pm 2c, \; c = \sqrt{h + \epsilon\eta}. \quad (7)$$

Characteristics played an important role in the early analytic work on runup of breaking waves[41,87] and are used in some numerical runup models as well[72,101,102]. In these references the technique is extended to two horizontal directions by operator splitting[102] or a partial characteristic description[72]. At the shoreline c becomes zero and both the characteristic equations and variables in (7) coalesce. Alternative equations or extrapolation must then be employed in a numeric model. Moreover, as long as $\partial(h + \epsilon\eta)/\partial x$ is nonzero at the shoreline, the spatial derivative of c is non-finite, which is a challenge from a computational point of view. Still, as demonstrated in the references runup can be treated accurately and efficiently by (7).

If the two-dimensional NLSW equations are linearized, we may eliminate the velocity to obtain a very simple equation

$$\eta_{tt} - (h\eta_x)_x = 0. \quad (8)$$

It might seem counter-intuitive for such an apparently nonlinear phenomenon, but linear equations work surprisingly well on a number of runup problems.

2.3. Lagrangian Coordinates and ALE Formulations

Problems involving material boundaries that are moving make application of Lagrangian coordinates tempting. In the limit $\mu \to 0$ the horizontal velocity is uniform in depth and it displays only a weak vertical variation for small μ. A Lagrangian description may thus be based on material columns of water for the NLSW equations and averaged columns for Boussinesq equations. A common choice for coordinates is the initial positions of the columns, but every mapping of these are feasible and do not lead to more complicated equations. The transformation between Eulerian and Lagrangian coordinates, (a,b), expresses that the latter are constant for a fluid particle; $a_t + \mathbf{v}\cdot\nabla a = 0$ and $b_t + \mathbf{v}\cdot\nabla b = 0$. Applying this transformation to equations of the form (2,3), with a value of z_α corresponding to depth averaged velocities, we arrive at equations for x, y and η according to

$$(h+\epsilon\eta)\frac{\partial(x,y)}{\partial(a,b)} = F(a,b), \qquad (9)$$

$$\frac{\partial(x,y)}{\partial(a,b)}\frac{x_{tt}}{\epsilon} = -\frac{\partial(\eta,y)}{\partial(a,b)} + O(\mu^2), \quad \frac{\partial(x,y)}{\partial(a,b)}\frac{y_{tt}}{\epsilon} = -\frac{\partial(x,\eta)}{\partial(a,b)} + O(\mu^2), \qquad (10)$$

where only the NLSW terms are spelled out and the Jacobi determinants are defined according to

$$\frac{\partial(x,y)}{\partial(a,b)} = \frac{\partial x}{\partial a}\frac{\partial y}{\partial b} - \frac{\partial x}{\partial b}\frac{\partial y}{\partial a}.$$

We note that time derivatives of positions are of order ϵ such that the ordering of (10) is consistent. Both a standard form of the Boussinesq equations[115] and a fully nonlinear version for plane waves[34,75] are available. Comparing (9,10) to the Eulerian counterpart we observe that the convective terms have vanished and that ungracious Jacobi determinants have appeared. Moreover, the continuity equation has been integrated once in time to directly express the volume conservation in a material fluid column, with F representing volume per area in the a,b plane.

For wave propagation in one horizontal direction the choice of marker coordinates is unproblematic, but non-rectangular two-dimensional horizontal domains have to be resolved with curvilinear grids or finite elements. For runup the Lagrangian description has the major advantage that the shoreline is located at temporally fixed values of a and b; the computational domain is independent of time. As a bonus this sometimes extends also to piston type wave paddles that may be cumbersome to incorporate in Eulerian models.

To the knowledge of the author no Lagrangian version of improved Boussinesq equations akin to (2,3) have been reported. Presumably, there are no principal difficulties in the transformation, but it will result in very lengthy expressions.

The fully Lagrangian description, where particles are traced throughout the fluid, is in fact an over-treatment of the moving boundary problem. We need only to follow the boundary particles and may thus employ a wider class of transformations, that may be referred to as Arbitrary Lagrangian

Eulerian (ALE). There is a large diversity with respect to selection of coordinates and dependent variables. In the simplest case with runup of plane waves an obvious choice is a time dependent stretching of the spatial coordinate in accordance with the shoreline motion. For simple three-dimensional geometries, typically having a single, moderately curved shoreline, a similar stretching with an along-beach variation may be employed[72,84]. With more complex geometries the transformation to time dependent curvilinear coordinates may be found by optimizing with respect to local refinement, variability and skewness of the grid (measured in the physical plane), while fixed grid-boundaries (in the coordinate plane) are constrained to follow the shoreline(s)[88].

In three-dimensional runup problems the ALE technique may be formulated more robust with respect to interior grid deformations due to currents than the purely Lagrangian ones. However, both traditional Lagrangian and ALE methods are susceptible to grid deformation when large inundation takes place in complex geometries. In addition bathymetric data must be interpolated at each time step in numerical simulations.

2.4. *The Shoreline*

We observe that the linear wave equation (8) has a singularity for $h = 0$. Provided that the gradient of h is finite at the shoreline both singular and regular solutions exist. The simplest example is that of a linear standing wave with frequency ω on an inclined plane defined by $h = \alpha x$. The regular solution, that has physical significance, then reads

$$\eta = A \mathrm{J}_0(2\omega\sqrt{\alpha x})\cos(\omega t + \delta), \qquad (11)$$

where A, δ are constants and J_0 is the Bessel function of zero order that may be expanded in even powers of its argument. For the singular solution of (8) we have $\eta \sim \ln(x)$ for small x, while the horizontal velocity has a pole of order 1. As a consequence there is a volume source at the singularity. Hence, runup models that do not conserve volume at the shoreline should be checked extra carefully for accuracy and spurious behavior. The mathematical structure with regular and singular solutions carries over to nonlinear equations as well as full potential theory[33] and is further discussed in a long wave context by Meyer[65,66] and others.

For nonlinear equations the shoreline is signified by $H = h + \epsilon\eta = 0$. In addition the position of the shoreline must be traced. We confine the discussion to two dimensions and assume that the shoreline always consists of the same material particle. The shoreline position, $x = \xi$, then moves according to $\frac{d\xi}{dt} = u(\xi, t)$. In Lagrangian description this is only an instance of the unknown $x(a, t)$, but in Eulerian models the equation for ξ must somehow be integrated. This is often achieved implicitly by filling and emptying of grid cells. For very steep and breaking waves, outside the long wave regime, the shore may not always correspond the same particle(s). Moreover, the waterline is affected both by bottom roughness, viscosity and capillary effects. The latter are generally more significant on the laboratory scale than for real tsunamis and storm surges, say. One should alway bear this in mind when models are calibrated against experiments.

2.5. *Additional Effects*

For waves propagating across oceans the rotation of the Earth must be taken into account as the Coriolis force. However, the local runup dynamics on a shore is hardly influenced by such effects and further discussion of this topic is thus omitted.

Bottom friction is often included in long wave models as a square of the velocity. A common form is

$$\mathbf{N} = -\frac{\epsilon}{\mu}\frac{K}{H}|\mathbf{v}|\mathbf{v}, \qquad (12)$$

where K is an empiric constant. Unfortunately, much of the systematic empiric data for bottom friction stem from steady channel flow and may not be relevant for the thin and rapidly changing fluid layers in runup. Moreover, there is no consensus concerning the use of drag terms like (12). Zelt and Raichlen (1991)[116] calibrated the formula against experiments[94] and then settled for $K = 5 \times 10^{-5}$. Comparing with the same dataset, Lynett et al. (2002)[57] suggested a value an order of 100 larger. We also observe that the above formula in principle becomes singular at the shoreline.

Bores may be included in the NLSW description as discontinuities in the velocity and the surface elevation. In numeric models these may be reproduced by special shock capturing schemes, generally applying approximate solvers of Riemann problems (evolution of step distributions). Alternatives that give finite width bores and may be used also with Boussinesq type equations are the roller model[62] or inclusion of a diffusion term, $\mathbf{E}$, in (3). For plane wave this term typically reads

$$E^{(x)} = H^{-1}(\nu(Hu)_x)_x. \qquad (13)$$

Such terms have often been included to stabilize numerical models, under cover of being a physical eddy viscosity effect. With a constant ν (13) is quite inefficient as a tool for controlling wave breaking. Zelt (1991)[114] made ν dependent on the velocity gradient in a similar term. ν was zero until the gradient reached a threshold value and then increased with the gradient beyond this. The critical value of the velocity gradient was related to the highest stable solitary wave, which has an amplitude 0.78 times the depth[b]. Kennedy et al.[42] expressed ν by η_t and pursued the idea further by including an explicit time relaxation for ν after activation of the diffusion. This method has been demonstrated to reproduce bores from experiments very well[42,57]. However, as pointed out in a reference[56], this procedure may not recognize a nearly stationary bore on a current, like the one that often forms during drawdown at beaches.

2.6. Numerical Modeling of Long Waves

There are numerous papers on numerical solution of NLSW and standard Boussinesq equations in general. Finite difference (FD) techniques dominate, but quite a number of finite element (FE) methods are reported as well. In traditional ocean and tsunami models the NLSW equations are often solved by explicit methods and discretized on staggered grids[64] (see Fig. 2), denoted as C-grid, B-grid etc. Many FD models employ discretization in finite volumes with fluxes defined at the interfaces to assure conservation of mass and momentum. This is particularly important for shock propagation models. Under-resolved models inherit a strong artificial dispersion that occasionally may mimic true dispersion. However, this cannot be accounted upon as a reliable feature and should be avoided. Grid sensitivity tests are crucial in this context.

Boussinesq equations must be simulated by implicit methods and lead to more CPU-time consuming models than the NLSW counterparts. As compared to standard Boussinesq formulations the additional higher order derivatives of the formulation (2,3) lead to extended computational molecules and more difficult boundary conditions. This, in addition to the lengthy expressions, render the implementation of this more general formulation somewhat deterring. However, two FD codes for numerical solution of this set may be downloaded from the Internet, namely the FUNWAVE[44,45]

---

[b]Recent investigations indicate that cross-wise instability appears at smaller amplitudes[40].

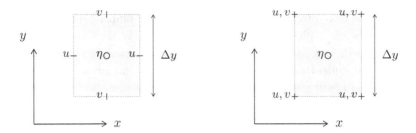

Fig. 2. Left panel: The C-grid that defines a basic configuration for mass conserving FD methods. It is used also for many Navier-Stokes models with η replaced by pressure. Right panel: The less used B-grid. This is well suited for representation of deformation, as needed in Lagrangian models. u and v are the velocity compenents in the x- and y-directions, respectively.

and COULWAVE[56] models. These contain a number of features, including runup facilities. Both models employ fourth order (five point) and second order differences for the $O(\mu^0)$ and the $O(\mu^2)$ terms, respectively, combined with higher order time integration. As a consequence, the leading dispersion in the wave celerity, namely the second term in (1), is kept clean of discretization errors, which is a natural objective for numerical solution of enhanced Boussinesq equations.

Solution of (2,3) by a finite element method is reported by Woo and Liu[108], where the problem with higher order spatial derivatives is resolved by introduction of extra dependent variables.

3. The Runup Phenomenon; Reference Solutions and Experiments

Runup is generally part of a more complete problem, where some incident wave is present in finite depth. This then undergoes shoaling and amplification, and may turn into a bore, before it reaches the beach and starts to run up (or down), often as a long and thin tongue that is retarded by gravity. The withdrawal phase is then characterized by the run-down of a thin layer of fluid, often with a strong bore, oriented onshore, evolving at the toe. This whole sequence involves many stages and different physical mechanisms. There is a rich literature on both shoaling and surf-zone dynamics that is outside the scope of this paper. Turbulence and sedimentation in the swash zone[12,54,83,85] are directly related to runup, but are also side issues in the present context.

3.1. The Non-Breaking Regime

The closed form solutions for runup are mostly associated with hydrostatic ($\mu \to 0$) theory. We will discuss some aspects of these solutions with reference to a wave in a generic wave tank, as depicted in Fig. 3, with an incident wave characterized by a length λ and amplitude A. The quantity ℓ is the propagation distance in constant depth, for instance the distance from a wave maker to the slope. In the limit $\mu \to 0$ the equations are independent of the choice for the scaling factor L, introduced at the start of Sec. 2. Rescaling with the length of the slope, $d \cot \theta$, as L we then realize that the solution of the problem in Fig. 3 is governed by the parameters $\epsilon = A/d$, $\kappa = L/\lambda$ and ℓ/λ.

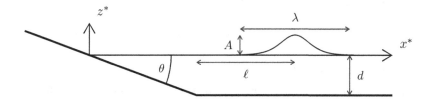

Fig. 3. Definition sketch of generic wave tank.

In 1958 Carrier and Greenspan[5] presented their much celebrated nonlinear, analytic solutions of the NLSW equations for runup on an inclined plane. They invoked a transformation that in dimensionless form reads

$$\sigma = 4\sqrt{x + \epsilon\eta}, \quad \lambda = 2(t - \epsilon u), \quad u = \frac{\phi_\sigma}{\sigma}, \quad \eta = -\frac{\phi_\sigma}{4} - \epsilon\frac{u^2}{2}. \tag{14}$$

The potential, ϕ, then solves the linear wave equation

$$\phi_{\lambda\lambda} - (\sigma\phi_\sigma)_\sigma = 0.$$

Among their explicit solutions were the nonlinear generalization of (11), the standing wave, as well as an initial value problem for runup. We will subsequently refer to these solutions as $CG\text{-}s^5$ and $CG\text{-}i^5$, respectively. A kind of breaking, in the sense that η becomes multivalued as function of x, is also contained in the standing wave solution. A major problem with the transformation is the specification of an incident wave. Carrier (1966)[4] suggested the coupling of the nonlinear solution with a linear outer solution, possibly including dispersion and a general bottom profile, through linear patching conditions. This idea has later been exploited by others, among which the work of Synolakis (1987)[94] has made most impact. Before going into details we will remark on some general properties of this approach. Since the transformed equations, as well as far-field and patching are linear, the solutions in σ and λ equal linear solutions in the physical plane with $\sigma^2/16$ and $\lambda/2$ substituted for x and t, respectively. Nonlinearity enters first through the transformation (14) back to the physical variables. When u is zero, as at the point of maximum excursion of the shoreline, the nonlinearity also disappears from this transformation. Hence, the maximum runup height for the combined linear/non-linear theory will be as for a purely linear one. Moreover, ℓ only gives a time shift when the constant depth region is governed by the LSW equations. For the wave tank problem in Fig. 3 the maximum runup height then becomes

$$R = AF(\kappa), \qquad (15)$$

where F is a function depending on the actual shape of the incident wave (an example is given below).

Solitary waves[68] are much used in theoretical runup investigations, even though they are more rarely seen in nature. They have permanent form, finite width and are easy to create in a laboratory. To leading order in μ and ϵ the solitary wave solution in dimensional form reads

$$\eta^* = A\operatorname{Sec}^2(k(x^* - ct^*)), \quad k = \frac{1}{d}\sqrt{\frac{3A}{4d}}, \quad c = \sqrt{gd}\left(1 + \frac{1}{2}\frac{A}{d}\right). \qquad (16)$$

In reality solitary waves are both nonlinear and dispersive, but the expression (16) may anyhow be used as initial condition in the combined linear/nonlinear calculations described above. Synolakis[94] applied Fourier transforms to find a simple asymptotic formula for the maximum runup height of solitary waves. Defining λ as $1/k$ we make the identification $\kappa = kd\cot\theta$ and write the formula on the form (15)

$$R = 3.042 A\kappa^{\frac{1}{2}} = 2.831 A(\cot\theta)^{\frac{1}{2}}\left(\frac{A}{d}\right)^{\frac{1}{4}} \quad \text{for} \quad \kappa \to \infty. \qquad (17)$$

When A and k in (16) are regarded as independent we observe that the runup decreases with the wavelength. This is general feature of the present theory that is found also for periodic waves. There are two main sources of errors in (17), namely the asymptotic assumption and the lack of dispersion on the slope. To check the effect of the first we may simply find R by solving (8) numerically. Application of a five point centered difference scheme, with

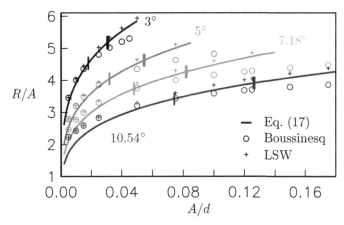

Fig. 4. Shallow water and Boussinesq predictions for maximum runup heights of solitary waves for different beach inclinations as indicated on the curves. The thick and thin columns represent breaking limits, (20), during runup and drawdown, respectively.

the closest node half a grid increment from the shore to implement the condition of zero flux, makes this a trivial task. The results are subsequently shown in Fig. 4.

The effect of dispersion on solitary wave runup is frequently overlooked when NLSW models are compared to experiments or dispersive models are compared with analytic NLSW solutions. This makes the model assessments imprecise, with a negative bias concerning model performance. We may employ optics to obtain a simple estimate on the variation of dispersion with depth. Assuming that the period is unchanged during shoaling, we find that the wavelength is reduced as $h^{-\frac{1}{2}}$ and the relative size of the leading dispersion term in the wave celerity vanishes proportional to h near shore. However, this result will not be valid in the proximity of the shore and nonlinear effects will lead to steepening of the wave front, which in turn enhance dispersion. Therefore, we resort to employment of a Lagrangian Boussinesq type model[34,76] (Sec. 4.1), that was readily at hand for the author. Converged results of both the LSW and Boussinesq models are compared to (17) in Fig. 4, that also contains the breaking criteria as given in Eq. (20). Shallow water theory over-predicts runup slightly for the larger A/d, but is surprisingly close to the Boussinesq theory even beyond the hydrostatic breaking limit. For large θ and small A/d, meaning small κ, the numeric solutions coincide, while the asymptotic formula (17) deviates. Hence, we clearly observe the asymptotic nature of the closed form solution

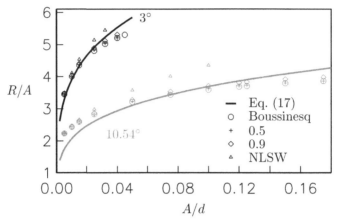

Fig. 5. Maximum runup heights for solitary waves. Results from combined Boussinesq/NLSW theory are marked by the transition depth $h(a_c)$ at equilibrium.

and that the agreement is good down to $\kappa = 1$, say, that corresponds to $R/A \approx 3.04$ in the figure.

We have also performed some numerical experiments to unravel where dispersion effects are important. To this end we introduce a variable $\mu(a)$, where a corresponds to the initial distance from the shore (Sec. 2.3). In deep water μ has its proper value, but over an interval $(a_c, 1.1 \times a_c)$ it is reduced to zero. As shown in Fig. 5, omission of dispersion over half the slope does not influence the maximum runup. Even if the hydrostatic region is increased to nearly the whole slope, the change in R is very small, though breaking occurs for a smaller A. The largest error occur when also the uniform depth region is made non-dispersive (data marked NLSW in the figure). This deviation depends strongly on ℓ that in the present simulation is chosen as $\ell = -\ln(0.00025)d/\sqrt{3A/d}$, which corresponds to an initial η^* equal to $A/1000$ at the start of the slope. In Fig. 6 we have compared the particle acceleration in the Boussinesq model to the contribution from the hydrostatic part of the pressure gradient, given by $\partial \eta/\partial x$, for a moderately steep incident wave. In finite depth there is a mild influence of the non-hydrostatic pressure (right panel). When the wave have just reached the shore the surface has an inclination of $36°$, which is steeper than an extreme (maximum amplitude) Stokes or solitary wave. Still, the non-hydrostatic part of the pressure gradient is noticeable only in a small region close to the shore.

Naturally, here we have merely scratched the surface concerning applicability of long wave theories. Still, it is indicated that for non-breaking waves

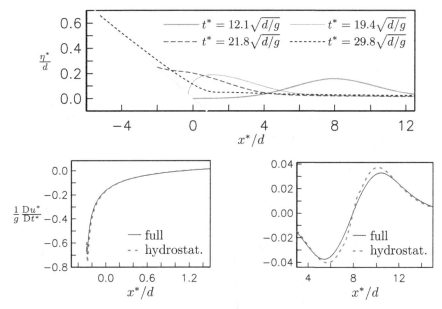

Fig. 6. Runup of a solitary wave with $A/d = 0.15$ and $\theta = 7.18°$. Upper panel: Surfaces for the incident wave, an early stage in runup, an intermediate stage and maximum runup. Lower panels: Comparison of the full acceleration to the part that stems from the hydrostatic pressure gradient. Left: Vicinity of beach for the second time (thin solid line in upper panel). Right: The incident wave (rightmost wave profile in upper panel).

the NLSW equation will often do for the near-shore region and runup. An equally important lesson is that dispersion should not readily be ignored for the constant depth propagation in simulations of wave tank experiments.

There are other analytic runup solutions that deserve mentioning. Some are related to those above, but involve different initial conditions, slides or modified geometries[3,6,39,49,74,77,92,95,97,103].

Another class of solutions contains the eigenoscillations in basins of parabolic bottom shapes that are summarized by Thacker[100]. There exist closed form solutions of the NLSW equations for the two lowest eigenmodes. In the first mode the whole fluid body moves back and forth in a uniform horizontal motion, driven by a uniform pressure gradient associated with a linear surface. Setting d equal to the maximum depth, identifying the half-width of the of equilibrium surface with L and choosing ϵd as the amplitude of the lowest harmonic in η^*, we obtain simple formulas for the 2D case

$$\eta = \cos(\sqrt{2}t)x - \frac{\epsilon}{4}\cos^2(\sqrt{2}t), \quad u = -\frac{1}{\sqrt{2}}\sin(\sqrt{2}t). \quad (18)$$

It is noteworthy that the eigenfrequency is independent of the amplitude. The second mode inherits a linear spatial variation of the velocity and a second order polynomial for the surface. The first eigenmode is much used for comparison with and validation of numerical methods. However, this may be a deceiving test due to the extremely simple spatial structure of the flow that is reproduced exactly by many methods that may give large discretization errors under other circumstances. It will serve best for validation of code or performance check of crude shoreline approximations. The second mode has somewhat better perspectives for assessment of runup models, but is used less often.

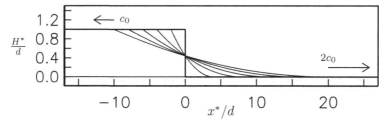

Fig. 7. The dambreak solution according to (19). Surfaces are depicted for $t^* = 0$, Δt^* .., where $\Delta t^* = 2d/c_0$.

An important analytic solution of the shallow water equations is that of dam-break[93], where a mound of water, usually step sized, is released from rest at an horizontal plane. The initial acceleration distribution and early development of the flow pattern are also extended to full potential theory[63]. Even though the initial water front is vertical, a smooth thin tongue will evolve. Assuming an initial condition with fluid height d for $x^* < 0$ and a dry bed for $x^* > 0$ we obtain a self-similar dilution wave from the NLSW equations. For $-c_0 t^* < x^* < 2c_0 t^*$ the solution reads (see Fig. 7)

$$H^* = \frac{d}{9}\left(\frac{x^*}{c_0 t^*} - 2\right)^2, \quad u^* = \frac{2}{3}\left(c_0 + \frac{x^*}{t^*}\right), \qquad (19)$$

where H^* is the total water depth and $c_0 = \sqrt{gd}$ is the long wave speed in depth d. Dam-break is a standard test case for Navier-Stokes models, and is quite common for long wave solvers as well. NLSW models may strive to reproduce the break at the limit to quiescent water, the huge accelerations for small times, and perhaps the thin and rapidly moving fluid tip. Otherwise, the solution (19) share the weakness of the eigenoscillations in parabolic basins concerning verification of NLSW models. The spatial structure of the flow is too simple to put many techniques to a proper test.

3.2. Transition to Breaking

Relevant overviews and background information on wave breaking may be found in, for instance, Cokelet[11], Peregrine[80] and, Synolakis and Skjelbreia[96].

From the NLSW solutions of the preceding section we also have breaking criteria for solitary waves during runup and drawdown on a plane beach on the form

$$\frac{A}{d} < C(\tan\theta)^{\frac{10}{9}} \quad \text{or} \quad \kappa^{\frac{5}{2}}\frac{A}{d} < \left(\frac{3}{4}\right)^{\frac{5}{4}} C^{\frac{9}{4}}. \tag{20}$$

For runup the constant $C = 0.82$ is reported[94], while $C = 0.48$ is obtained for drawdown[15]. These findings are consistent with the commonly observed feature that waves break more easily during withdrawal. The breaking limits are depicted in Fig. 4. From the figure we observe that the breaking criterion for runup is too strict compared to the Boussinesq results. This is as expected since dispersion slow down the steepening during shoaling. Grilli et al.[24] have investigated breaking of solitary waves on a beach by a boundary integral method for full potential theory. The reference recognized two breaking regimes: one with breaking before the shoreline, or on-shore plunging, and an intermediate regime with a (nearly) vertical front at, and a little beyond the equilibrium shoreline, which transforms into a smooth swash without any appreciable plunging or spilling. Applying curve fitting to the breaking/no-breaking domains in the A, θ plane, approximate criteria were presented as

$$\frac{A}{d} = B\tan^2\theta, \tag{21}$$

where $B = 16.9$ and $B = 25.7$ correspond to the lower boundaries of the intermediate and strongly breaking regimes, respectively. The criterion (21) is generally much more relaxed than (20). However, it must be remarked that the definition of breaking is ambiguous and that (21) is based on few data points, in particular for small A/d and θ where (20) is most likely to apply. Experiments[34] have confirmed the existence of the intermediate zone and even revealed waves that were overhanging at the shore, but still did not proceed to breaking. Instead, the steepening process was reversed and a smooth swash emerged from the toe of the wave.

Plunging breakers are outside the long wave regime, but fully developed bores may be crudely presented as discontinuities in shallow water solutions. This theme is treated in Chapter 2 (by LeVeque and George) in this volume and we will merely attach a few comments on the runup aspects. The most remarkable feature for bore runup[28,41,67,87] is that the bore collapses rapidly at the shoreline, giving birth to a long thin runup tongue with a height that is proportional to the square of the distance from the instantaneous shoreline. The swash is dominated by gravity and the runup trajectory becomes a parabola in the x,t plane. This transition from a steep incident wave to a swash is closely related to dam-break[34,67,81] (see Fig. 7). The runup of steep non-breaking waves and finite width bores are quite similar to that of the idealized bore solution; a steep front is transformed into a thin swash zone, where gravity often dominates over pressure effects[34,114]. Moreover, there are large accelerations and velocities in the early phases of runup, when the steep front vanishes. Small timing errors in experiments or models may lead to large deviations temporarily, while quantities like runup height and overall pattern is less influenced. This may cause an under-rating of model performance. The occurrence of a very thin runup layer that even may vanish as the square of the distance from the tip, seemingly argue against the application of a linear onshore extrapolation of the surface that is employed in some models. Even more so, perhaps, since the rundown phase display even thinner layers. On the other hand, bottom roughness, finite width bores and turbulence will modify the swash zone dynamics. Since, in addition, the feedback of a thin swash to the main body of fluid may be small, it is quite likely that surface extrapolation may be acceptable also in such cases.

3.3. *Experiments*

There is only a limited number of published experimental investigations that are commonly used for assessment of theoretic models. Hall and Watts (1953)[26] measured runup heights of solitary waves for different amplitudes and beach inclinations. Synolakis (1987)[94] published runup heights of breaking and non-breaking solitary waves on a 1.0 in 19.85 slope and amplitudes in the range $0.004 < A/d < 0.6$. In addition, data from a set of wave gauges, on and off the beach, were combined to rather complete surface elevations at different times. These surfaces have been valuable for validation of numerical models, in particular for the breaking cases. One such data-set is displayed in Fig. 10.

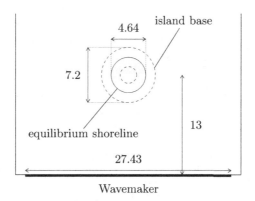

Fig. 8. Definition sketch of conical island experiment, with scales in meter. The outer dashed circle is the base of the island, the fully drawn circle is the equilibrium shoreline when $d = 0.32$m.

Briggs et al. (1995)[2] investigated runup on a conical island, with slope 1 in 4, situated on plane bed of equilibrium depth, d, 0.32m or 0.42m (Figs. 8 and 9). This experiment was inspired by observed runup patterns at a rounded island for the Flores tsunami in 1992[112]. Among the incident waves were plane solitary waves with amplitudes a little short of $A/d = 0.05, 0.1, 0.2$. Wave heights were measured at different offshore wave gauges and special gauges at the sloping bottom provided coarse digital series of inundation and withdrawal around the island. The experiments are also described in detail by Liu et al. (1995)[52] with the first numeric simulations on this problem. It is noteworthy that the breaking criterion (21) states that no solitary wave will break during runup on a plane 1 in 4 beach. Solitary waves of amplitude 0.2, or less, may then come close to breaking[c] only during withdrawal or due to three-dimensional effects. The criterion (20) yields $A/d = 0.175$, but the slope is too steep and the solitary wave too high (and short) for shallow water theory to apply. The most remarkable feature of the measured runup was a pronounced local maximum at the rear of the island, due to interaction of waves coming from either side. It is noteworthy that linear theories[14,38] reproduce the experiments very well, save for the lee side and then particularly the local maximum for the intermediate amplitude $A/d = 0.1$. The conical island experiment was used as a benchmark problem for the Second workshop on long wave runup[113].

---

[c]The Boussinesq model[34] employed in Sec. 3.1 indicates breaking during retreat for $A/d = 0.2$, but not for the smaller amplitudes.

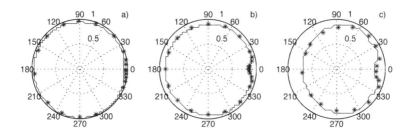

Fig. 9. Measured[2] (stars) runup on conical island with recent simulations by Lynett et al.[57] (lines). (a): $A/d = 0.05$, (b): $A/d = 0.1$, (c): $A/d = 0.2$. Reproduced from Lynett et al. by courtesy of the Coastal Engineering.

There are other experimental investigations of runup that, so far, are less used for verification of long wave models[34,49,50,67,69,111].

4. Runup Models

We may recognize at least three key problems in runup modeling:

(1) The shoreline position is unknown *a priori*, it must be determined as a part of the solution procedure. The computational domain must somehow be adjusted to the shoreline motion.
(2) The singularity at the shoreline means that there do exist singular solutions of the PDEs that may be exited by numeric errors.
(3) It is difficult to find a sufficient number of accurate boundary conditions, in contrast to, for instance, the symmetry condition at a vertical wall. In finite difference methods one sided representations are generally required at shoreline. These may be implemented directly or indirectly by fictitious points combined with extrapolation. Common consequences are that the order of accuracy of the numerical method is reduced at the waterline and a production of noise that must be controlled by filtering or dissipation.

In view of these points the performance of runup models should be carefully assessed. Results should be compared to the analytic solutions or experiments on runup, or at least compared to computational results obtained independently. In addition, grid dependence in the solution, which always can be investigated, should be reported. It is surprising and disappointing how seldom the latter is found in the literature.

Some of the literature that addresses moving shorelines is focused on storm surges and tides. Such applications involve smaller gradients and longer time scales than runup of tsunamis, or swells, and are thus less challenging. The contributions of this kind often omit both testing and detailed descriptions of shoreline dynamics. Hence, it is difficult to evaluate the potential of such published methods concerning runup of steeper waves.

In the subsequent review, priority will be given to articles where validation in at least one of the above fashions is documented. The methods are grouped in two: those that remove the time dependence of the computational domain by a transformation, and those formulated on a fixed grid, possibly with some auxiliary nodes at the shoreline. Both groups contain finite difference as well as finite element methods. The reader should always bear in mind that the referenced articles vary in style and objective. Hence, when we are enabled to refer discussions on errors and shortcomings from a particular work this may equally well reflect quality and open-mindedness, on the authors side, as inferior model performance relative to other publications.

4.1. *Transformed Domain Methods*

Application of a Lagrangian, or an arbitrary Eulerian-Lagrangian (ALE), description generally eliminates the tracing of an *a priori* unknown shoreline (point 1 above). Several runup models of this kind have been reported, even though they are in minority relative to the Eulerian counterparts.

An early discussion of runup and Lagrangian coordinates is given by Shuto (1967)[89], but for the linear shallow water equations only. Hence, the potential of the Lagrangian description was not really exploited. Later Goto (1979)[20] and Goto and Shuto (1980, 1983)[21,22] compared 2D numeric results with linear solutions and the nonlinear standing wave solution CG-s^5. The description of the method was sketchy, but the agreement with the analytic solution was very good and markedly better than for some early Eulerian models. In 1983 Pedersen and Gjevik[75] applied Lagrangian coordinates to a 2D Boussinesq set with standard linear dispersion properties. Employing finite differences on a staggered grid the authors reproduced analytical solutions[92] accurately. Computed runup of non-breaking solitary waves were compared to experiments, with good agreement for moderately steep slopes. The model has later been revisited and improved by Jensen *et al.* (2003)[34], who also discussed the significance of different $O(\epsilon\mu^2)$ terms in the runup. A 3D Lagrangian FD model was reported by Johnsgard and

Pedersen (1997)[37] and Johnsgard (1999)[36] with applications to runup and wave generation by sub-aerial slides in idealized lake geometries. The model was based on a B-grid and verified by extensive grid refinement and comparison with CG-s^5 and oscillations in a parabolic basin[100].

Lynch and Gray (1980)[55] described a finite element technique with time dependent shape functions and a grid that could be adapted to a moving shoreline. However, their model was not tested on any standard problem. The first runup model that inherited both three-dimensionality and dispersion was published by Zelt and Raichlen in 1990[115]. They employed a finite element technique with a partition in quadrilaterals in the Lagrangian coordinates. One appealing feature of this kind of formulation is the so-called natural boundary condition for the shoreline. After reproducing CG-s^5 accurately the authors applied the model to the response of an idealized harbour to incident swells. Their equations were similar to the standard Boussinesq equations, apart from some extra error terms of order ϵ^2 (see Sec. 2) that appeared for curved beach profiles. This minor fault was eliminated in the succeeding work[114], where wave inundation of a horizontal shelf were modeled with a bottom friction term, giving good agreement with experiments. In a subsequent paper[114] Zelt included wave breaking by means of a diffusion (Sec. 2.5) dependent on the local wave steepness, or, rather, the compression rate of the Lagrangian water columns. Computations were compared with non-breaking ($A/d = 0.04$) and breaking ($A/d = 0.28$) solitary wave runup experiments[94]. Excellent agreement was observed for non-breaking runup, while there were discrepancies for breaking waves. In particular, a spurious, nearly vertical front persisted during inland excursion, which presumably was due to the diffusion term. Comparison was also made between Boussinesq, as well as NLSW, simulations and experiments[19] of solitary wave runup on a steep ($\theta \approx 20°$) slope. It is noteworthy that the Boussinesq results reproduced the experiments much closer (see discussion in Sec. 5). Later another Lagrangian FE method for the NLSW equations was reported by Petera and Nassehi (1996)[82]. Unfortunately, little documentation on the runup was given and there is no evidence of any improvement over the FE technique described above.

In two dimensions, the fluid domain can be mapped onto a fixed interval through a time dependent, but spatially uniform, coordinate stretching in accordance with the waterline motion. An example was given for the NLSW equations by Johns (1982)[35] who employed a staggered grid and linear extrapolation of η to obtain the shoreline position. A numeric solution showed fairly good agreement with an CG-s^5 solution with a very small amplitude. A similar technique was reported by Takeda (1984)[99], who used a semi-characteristic method to reduce the need for one-sided approximations at the shore. Good agreement with CG-s^5 was found for runup, while a somewhat degraded performance was observed for drawdown. The paper contains a general discussion on employment of asymmetric difference equations at the shore, but no model test that could support any firm conclusion was included. The approach of Johns[35] was generalized to three dimensions by Shi and Sun (1995)[88] who invoked an orthogonal transformation and then a C-grid discretization on a fixed coordinate domain that was composed of rectangles. The technique was applied to a real storm surge inundation. The large deformation of the grid imposed by the adaptation to a real geometry clearly called for grid-refinement studies, but none were presented.

In Özkan-Haller and Kirby (1997)[72], the NLSW equations were rewritten in a partial characteristic form and solved on a transformed domain of rectangular shape by spectral collocation. Chebyshev polynomials and trigonometric functions were used for the cross- and along-shore directions, respectively. The CG-s^5 solution was reproduced with high accuracy and the method was then applied to edge-wave instability. An advantage over fully Lagrangian techniques is that strong shear currents may be included without grid distortion. Recently, Prasad and Svendsen (2003)[84] have employed a similar transformation with a finite difference technique for the NLSW equations and obtained close agreement with both CG-s^5 and CG-i^5. The authors also presented a FD technique on a fixed grid, where they went to some length in describing the shoreline accurately by carefully discretized boundary conditions. Also this technique displayed good results, but was markedly more prone to shoreline fluctuations than the transformation method and displayed a slower convergence according to grid refinement tests. Zhou et al. (2004)[117] make the same simple transformation as Johns[35] for the two-dimensional problem. The NLSW equations are then solved by a mesh-less technique, utilizing smooth shape functions and collocation. The paper is brief and only dam-break is presented as verification.

4.2. *Fixed Grid Methods*

In a fixed grid the key problem is to include/exempt points and allot proper values to field variables at the boundary. For finite difference methods this is most commonly attempted by more or less ingenious extrapolation schemes. A number of such are reviewed in Imamura (1996)[32]. Balzano (1998)[1] tested a series of NLSW methods for tidal flooding, mainly based on the C-grid, on a small amplitude second eigenmode in a parabolic basin[100]. To a greater or lesser extent all techniques gave near-shore fluctuations. However, the degree of nonlinearity was too small to allow any inference on the general runup performance of the models.

Apart from simulations of flooding based on linear theory[86], the first proper runup model was reported by Sielecki and Wurtele (1970)[90]. They employed three different finite difference techniques, among them the Lax-Wendroff method, for the NLSW equations with the Coriolis term. A simple extrapolation combined with an asymmetric representation was used to follow the shoreline. Among the test cases were CG-i^5, that was well reproduced except for a small discrepancy at shoreline, and a three-dimensional eigenoscillation in a parabolic basin[100] that was simulated perfectly. In 1975 Flather and Heaps[13] published a 3D FD technique for the NLSW equations, based on the C-grid, with a scheme to update wet and dry regions. However, the model was applied to tidal oscillations in a bay without any verification of its runup performance. Hibberd and Peregrine (1979)[27] employed the Lax-Wendroff technique to the NLSW equations in conservative form and obtained a model for non-breaking waves as well as bores. According to the authors the simple shoreline method of Sielicki and Wurtele[90] did not suffice for these problems. After attempting various methods they settled for a predictor-corrector sequence of linear extrapolations and applications of the physical equations in the vicinity of the shoreline. Even though some near-shore deviations were identified, both CG-i^5 and CG-s^5 were reproduced well. Very close agreement were reported with analytical runup heights[67], while the theoretical values overshot measurements[69]. This was explained by viscous effects that must influence the thin swash zone significantly on laboratory scale. Later Kobayashi (1987)[46] adopted the method and applied it to runup on rough slopes.

Kowalik and Murty (1993)[47] employed linear onshore extrapolation of velocity and surface elevation with a C-grid discretization of the NLSW equations. Their method gave reasonably good results for CG-i^5 and the first mode in a parabolic basin, but with significant noise at the shoreline. Then they used the method to compute the inundation in a real tsunami event.

With the experiments on solitary wave runup on a conical island Briggs et al. (1995)[2] (Sec. 3.3) provided what still is the only 3D dataset well suited for verification of models. In a companion paper Liu et al.[52] presented NLSW simulations with a C-grid model with upwind representation of the convective terms. The moving shoreline was represented through a sequence of over-toppings and dry-outs in a step-bathymetry. In spite of some deviation the maximum runup heights were generally predicted very well for the smaller amplitudes. Moreover, even though the sharp local maximum at the lee side was not fully reproduced for $A/d = 0.1$, the nonlinear model fared much better than linear solutions[14,38] in this respect. Recently, the same method was re-applied to this case for $A/d = 0.1$, utilizing a hierarchical grid to obtain selective grid-refinement around the island[10]. However, no substantial improvement was obtained. The results of Liu *et al.* for the highest amplitude were less good. According to the discussion in Sec. 3.3, this case really requires a dispersive model. At the time the only well documented Boussinesq type model for three-dimensional runup was the Lagrangian method of Zelt and Raichlen[115], described in Sec. 4.1. Regretfully, to the knowledge of the author this finite element model is never applied to the conical island problem.

Titov and Synolakis (1995)[101] used the characteristic form of the 2D NLSW equations that were solved by a corrected leap-frog method allowing for non-uniform grids. This enabled the inclusion of an auxiliary shoreline point in the otherwise fixed grid. The position and velocity for the extra point were projected from the nearest wet point. Good agreement with experiments[94] and the formula (17) was obtained for non-breaking solitary wave runup. The method also gave rather good agreement with experiments on a breaking wave ($A/d = 0.3$), save for some pronounced deviation in the early phase of runup. However, this may partly be due to the sensitivity of the solutions to small time differences at this stage (Sec. 3.2), but omission of dispersion did probably contribute as well. Later the method was generalized to three dimensions by operator splitting[102]. Comparison was then made to the conical island experiment[2] with very good agreement for $A/d = 0.1$ and somewhat larger errors for $A/d = 0.2$. The latter may again be due to the absence of dispersion. Physical tsunami applications displayed encouraging results indeed and demonstrated that the method is robust in complex situations and handles features as merging of shorelines. It is incorporated in the tsunami model known as MOST[73].

Madsen *et al.*[62] reported a 3D Boussinesq technique with improved linear dispersion characteristics, based on depth integrated volume fluxes, and a surface roller model for breaking. The beach was made slightly permeable with a porosity, rapidly decreasing from unity to a small residual value beneath the surface. As a consequence the equations could be solved throughout a region that was extended beyond the shoreline, with no explicit reference to its position. However, as communicated by the authors, this technique was not without weaknesses. The dispersion terms was turned off in the onshore motion, noise had to filtered and there was a mass deficit in the swash tongue due to the porosity. The latter caused marked deviations from the CG-s^5 solution, while the overall agreement was still reasonably good. Subsequently, the porous shore technique were amended and incorporated in the FUNWAVE model[45] for improved Boussinesq equations (Sec. 2.6), with results presented in a series of papers. In a paper on rip-currents Chen et al. (1999)[8] interpreted the porosity as slots in the seabed and counterbalanced the volume loss by adjusting the inter-slot beach level. Wave breaking was represented by an eddy viscosity depending on the temporal surface gradient (Sec. 2.5). This model was put to test by Kennedy et al. (2000)[42] through comparison with experiments on breaking in finite depth and the analytic CG-s^5 solution. The model reproduced the breaking well, while the agreement with CG-s^5 was substantially improved relative to Madsen et al.[62], even if some deviation persisted. Chen et al. (2000)[9] did the conical island test. The agreement for the smaller amplitudes, $A/d = 0.05, 0.1$ was fairly good, but poorer rather than better relative to the pre-existing NLSW computations[52,102]. For the higher amplitude, $A/d = 0.2$, where dispersion is more important, the results improved, at least compared to Liu et al. (1995)[52]. Unfortunately, simulations based on the porous bed technique appeared not to have been compared to the detailed experimental surface profiles for breaking and non-breaking waves[94].

The runup facility in the COULWAVE model[56] for the improved Boussinesq equations was described and tested in Lynett et al. (2002)[57]. Again wet and dry nodes were recognized by the total water depth, while the discrete equations at the shoreline were based systematically on fictitious nodes beyond the shoreline. Linear extrapolation from neighboring points in the fluid was used to assign values to the onshore nodes. Higher order differences were then used throughout the computational domain. In principle, this is a simple procedure that may seem to depend heavily on the analytic nature of the solution. The authors presented a good description of the implications of the procedure for the computational molecules adjacent to the shore. In addition to the general loss of accuracy and noise (arrested by a 9 point filter) at the shoreline, the higher order spatial derivatives were annihilated, which imply omission of dispersion terms. Wave breaking was included in a similar manner as in Kennedy et al. (2000)[42]. Model results agreed well with CG-s^5 and the second eigenmode in a parabolic basin[100]

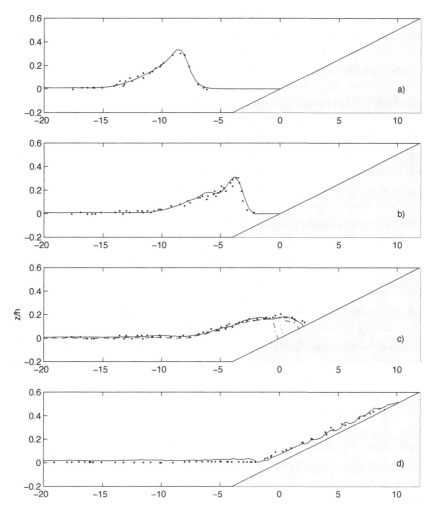

Fig. 10. Breaking solitary wave runup. Experiments from Synolakis[94] are indicated by markers, while the fully drawn lines are solutions from Lynett et al.[57]. Panel (c) includes also results from Zelt[114] (dots), Titov and Synolakis[101] (dash-dot) and the N-S computations of Lin et al.[51](dashes). Reproduced from Lynett et al. by courtesy of the Coastal Engineering.

was accurately reproduced. Also the conical island test was made with good results. For runup of a breaking wave ($A/d = 0.3$) impressive agreement was obtained with both experiments[94] and a turbulent Navier-Stokes simulation[51](Fig. 10). Even though this is unlikely to be the reason for the good outcome, it must be noted that the choice of the coefficient in the bottom-drag was optimized. As pointed out by the authors, there are still some unresolved shortcomings in the model for features like over-topping and reentry of a runup tongue into quiescent water[56].

Another line of runup modeling is based on the NLSW equations in conservative form, finite volume techniques and approximate Riemann solvers to include bores as sharp jumps in the solution. This branch will be covered in depth in Chapter 2 in this volume by LeVeque and George[48]. Hence, we refer briefly only a selection of contributions, with emphasis on runup tests. The shoreline may be treated by the special Riemann problem related to penetration of gas into vacuum or dam-break. However, this leads to overestimation of the pressure gradient and acceleration at the shoreline. Hence, modelers often resort to complete or partial exclusion of very shallow cells from the general computation scheme. The flux and source terms in the momentum equation (see below Eq. (6)) are generally separated by operator splitting, even though Watson *et al.* (1992)[106] suggested the inclusion of the source term in the Riemann problem by a local transformation to an accelerating frame of reference. Sleigh *et al.* (1998)[91] invoked a discretization into triangular volumes which enables flexible local grid refinement. The method was tested on various dam-break problems and bore-runup. Good agreement was observed with the analytical solution for the standard dam-break problem, save from small deviations at the toe. Two-dimensional simulations were presented by Hu *et al.* (2000)[30] including runup of non-breaking solitary waves, that agreed closely with an analytic solution[94], and the standard dam-break problem. The latter case demonstrated that exemption of shallow cells provide a better solution near the fluid tip than application of a modified Riemann solution. In Brocchini *et al.* (2001)[3] operator splitting was employed on a rectangular grid and a Riemann technique was used for the shoreline. Results compare well with the analytic solution CG-i[5] and the standard dam-break problem. However, for the latter there was noticeable spurious behavior close to the water line. The particular strength of Hubbard and Dodd (2002)[31] is a hierarchical mesh refinement, which means that finer rectangular grids may be superimposed on coarser ones in regions that require high resolution, as the shore and swash zone. Close agreement, save for small irregularities at the shore, was obtained with CG-s[5]. Surprisingly, the reproduction of the first nonlinear eigenmode in a parabolic basin was less good.

A common objective for this kind of models is to preserve sharp bore fronts. It would have been interesting to see a detailed comparison with real, finite width bores, as breaking solitary waves[94].

The finite element method has a much weaker merit in Eulerian runup modeling than the finite difference counterpart. Several methods for including a moving shoreline in a static FE grid have been proposed[105], but few have been validated by comparison to reference solutions or systematic grid refinement. An interesting attempt on solving a two-dimensional Boussinesq equation with the differentiable Hermitian cubic shape functions was made by Gopalakrishnan and Tung (1983)[18]. The pressure gradient was used directly to compute the shoreline acceleration which then was integrated to yield the position. The shore element was then modified at each time step, and split/deleted when appropriate. However, the only application was solitary wave runup, without any comparison to other solutions nor application of grid refinement tests. Moreover, the presented solutions clearly displayed noise and spurious features. Later the method was modified[17] to the three-dimensional NLSW equations and applied to tidal flooding, again with inadequate testing of the runup performance. Umetsu (1995)[104] employed partially dry triangular elements, where the velocities on dry near-shore nodes are imposed as averages of the values from the adjacent wet ones. The technique performed well for broken dam flow, but produced outspoken deviation from the CG-i[5], in particular close to the shore. In Takagi (1996)[98] the technique was applied to the conical island problem with reasonably good, but not convincing, result.

5. Discussion

The crucial point in runup modeling is the representation of the shoreline motion, for which the presence of a singularity (Sec. 2.4) calls for caution. The literature refer many rather crude as well as sophisticated methods for the moving boundary. Even though noise in the vicinity of the shoreline is often observed there are few severe problems that have been attributed to the singularity. It is tempting to assume that this problem is not that grave after all, or that a logarithmic singularity is too weak to be fatal. On the other hand, too many authors have treated numerical convergence rather loosely. More rigid grid refinement studies might reveal that there indeed are irregularities at the shore, even though they in most applications will be overshadowed by other error sources. Anyway, the bottom line is that there is a diversity of working runup models about.

The recent progress in long wave runup modeling may roughly be divided into two directions: one where the runup facility is integrated in general purpose wave propagation models with high order inherent dispersion and another where increasingly robust or accurate shoreline representations are included in NLSW models, with features like adaptive grids and shock capturing. Some of the NLSW models are both well tested, efficient, and flexible enough for, for instance, realistic case studies of tsunami, while the existing Boussinesq type models still may miss a little on runup/drawdown, bore presentation and computer efficiency for such applications. On the other hand, dispersion is important for wave propagation in finite depth, not only for swells, but also for many tsunamis. This leads to the questions; when are the different kind of long wave models appropriate and how may they be combined in applications that require the strengths of both kind of models?

Our little study of significance of dispersion, reflected in Fig. 5, suggests that the NLSW equations may be very accurate for the near-shore region and runup for non-breaking waves. Regretfully, it is difficult to attain further insight on this point from the literature. Truly, in some of the cases of non-breaking solitary wave runup, reviewed in the previous section, the inclusion of dispersion led to improved agreement with experiments or more general models[57,114]. However, the difference may be due to accumulation of dispersion effects in finite depth propagation and shoaling (Sec. 3.1) that change the incident wave before the shoreline is reached, and that are missed out in the NLSW solution. For the case of breaking solitary wave runup[94] the improved Boussinesq solution[57] is clearly superior to those of the NLSW[101] and standard Boussinesq equations[114] (Fig. 10). Still, again the effect of dispersion on the incident presumably is important. In addition the bore representation are very different in the respective models, and even somewhat dubious for the standard Boussinesq formulation in question. Hence, it is difficult to draw any firm conclusion on the local near-shore effect of non-hydrostatic pressure and more investigations are required, where local effects are emphasized.

If, for a given application, the dispersion can be neglected near-shore, but not elsewhere, nesting of models may be desirable. For instance, a higher order Boussinesq type description for deep water propagation and shoaling should be coupled with an efficient NLSW solver with shock handling and robust runup properties for the surf and swash regions. A one way coupling from Boussinesq to NLSW equations is principally straightforward. On the other hand, a two way coupling involves conceptual problems. The NLSW solver will produce and preserve shocks sharp, except at the shoreline. In finite depths the shocks may vanish only due to dissipation. Hence, the reflected waves from shallow region may be difficult to convey to a

Boussinesq type model without severe problems. To the knowledge of the author no such attempt on a complete two way coupling has been reported in the literature.

For Lagrangian, and curvilinear, models very good performance is reported for runup of waves in 2D or simple 3D geometries. This makes transformed domain models an excellent tool for academic studies. However, a widely accepted weakness of such methods is the strong grid deformation that may occur in demanding applications. No well documented attempt to apply such models to real tsunami studies has been reported. In fact, in recent years Lagrangian coordinates, in general, have been most popular in connection with smooth particle hydrodynamics that is a meshless method. It would have been interesting to see how well such a method could be employed with the NLSW equations and free boundaries.

Acknowledgments

The author wishes to thank Professor P. L.-F. Liu and Professor P. Lynett for their helpful assistance.

References

1. A. Balzano. Evaluation of methods for numerical simulation of wetting and drying in shallow water flow models. *Coast. Eng.* **34**, 83–107 (1998).
2. M. J. Briggs, C. E. Synolakis, G. S. Harkins and D. R. Green. Laboratory experiments of tsunami runup on circular island. *Pure and Applied Geophysics* **144**(3/4), 569–593 (1995).
3. M. Brocchini, R. Bernetti, A. Mancinelli and G. Albertini. An efficient solver for nearshore flows based on the WAF method. *Coast. Eng.* **43**, 105–129 (2001).
4. G. F. Carrier. Gravity waves on water of variable depth. *J. Fluid Mech.* **24**, 641–659 (1966).
5. G. F. Carrier and H. P. Greenspan. Water waves of finite amplitude on a sloping beach. *J. Fluid Mech.* **4**, 97–109 (1958).
6. G. F. Carrier, T. T. Wu and H. Yeh. Tsunami run-up and draw-down on a plane beach. *J. Fluid Mech.* **475**, 79–99 (2003).
7. Y. Chen and P. L.-F. Liu. Modified Boussinesq equations and associated parabolic models for water wave propagation. *J. Fluid Mech.* **288**, 351–381 (1995).
8. Q. Chen, R. A. Dalrymple, J. T. Kirby, A. B. Kennedy and M. C. Haller. Boussinesq modeling of a rip current system. *J. Geophys. Res.* **104**, 20617–20637 (1999).
9. Q. Chen, J. T. Kirby, R. A. Dalrymple, A. B. Kennedy and A. Chawla. Boussinesq modeling of wave transformation, breaking, and run-up. Part II: 2D. *J. Waterw., Port, Coast., Ocean Engrg.* **126**(1), 48–56 (2000).

10. Y.-S. Cho, K.-Y. Park and T.-H. Lin. Run-up heights of nearshore tsunamis based on quadtree grid system. *Ocean engineering* **31**, 1093–1109 (2004).
11. E. D. Cokelet. Breaking waves. *Nature* **267**, 769–774 (1977).
12. B. Elfrink and T. Baldock. Hydrodynamics and sediment transport in the swash zone: a review and perspectives. *Coastal Engineering* **45**, 149–167 (2002).
13. R. A. Flather and N. S. Heaps. Tidal computations for Morecambe Bay. *Geopshys. J.R. Astr. Soc.* **42**, 489–517 (1975).
14. K. Fujima. Application of linear theory to the computation of runup of solitary waves on a conical island. In *Long-wave runup models,* Eds. H. Yeh, C. E. Synolakis and P. L.-F. Liu (World Scientific Publishing Co., 1996), p. 221–230.
15. B. Gjevik and G. Pedersen. Run-up of long waves on an inclined plane. Preprint series in applied mathematics, Dept. of Mathematics, University of Oslo (1981).
16. M. F. Gobbi, J. T. Kirby and G. Wei. A fully nonlinear Boussinesq model for surface waves. Part 2. Extension to $O(kh)^4$. *J. Fluid Mech.* **405**, 181–210 (2000).
17. T. C. Gopalakrishnan. A moving boundary circulation model for regions with large tidal flats. *Int. J. Num. Meth. Eng.* **28**, 245–260 (1989).
18. T. C. Gopalakrishnan and C. C. Tung. Numerical analysis of a moving boundary problem in coastal hydrodynamics. *Int. J. Num. Meth. Fluids* **3**, 179–200 (1983).
19. D. G. Goring. Tsunamis: The propagation of long waves onto a shelf. Ph.D. thesis Rep. KH-R-38, Calif. Inst. Technol., Pasadena, California (2003).
20. C. Goto. Nonlinear equations of long waves in the Lagrangian description. *Coast. Eng. Japan* **22**, 1–9 (1979).
21. C. Goto and N. Shuto. Run-up of tsunamis by linear and nonlinear theories. In *Coastal Engineering. Proc. of the seventeenth Coastal Engineering Conf.,* vol. 1 (Sydney, Australia, 1980), p. 695–70.
22. C. Goto and N. Shuto, Numerical simulation of tsunami propagations and run-up. In *Tsunamis–Their Science and Engineering,* Eds. K. Lida and T. Iwasaki (TERRAPUB, 1983), p. 439–451.
23. S. Grilli and I. A. Svendsen. Computation of nonlinear wave kinematics during propagation and run-up on a slope. In *Water wave kinematics* (Kluwer Academic Publishers, 1990), p. 378–412.
24. S. T. Grilli, I. A. Svendsen and R. Subramanya. Breaking criterion and characteristics for solitary waves on slopes. *J. Waterw., Port, Coast., Ocean Engrg.* (1997).
25. S. Guignard, R. Marcer, V. Rey, K. Kharif and Frauníe. Solitary wave breaking on sloping beaches; 2-D two phase flow numerical simulation by SL-VOF method. *Eur. J. Mech. B-Fluids* **20**, 57–74 (2001).
26. J. V. Hall and J. W. Watts. Laboratory investigation of the vertical rise of solitary waves on impermeable slopes. Tech. Memo. 33, Beach Erosion Board, U.S. Army Corps of Engrs. (1953).

27. S. Hibberd and D. H. Peregrine. Surf and run-up on a beach: a uniform bore. *J. Fluid Mech* **95**, 323–345 (1979).
28. D. V. Ho, R. E. Meyer and M. C. Chen. Long surf. *J. Mar. Res.* **21**, 219–232 (1963).
29. S.-H. Hsiao, P. L.-F. Liu and Y. Chen. Nonlinear water waves propagating over a permeable bed. *Phil. Trans. R. Soc. Lond. A* **458**, 1291–1322 (2002).
30. K. Hu, C. G. Mingham and D. M. Causon. Numerical simulation of wave overtopping of coastal structures using the non-linear shallow water equations. *Coast. Eng.* **41**, 433–465 (2000).
31. M. E. Hubbard and N. Dodd. A 2D numerical model of wave run-up and overtopping. *Coast. Eng.* **47**, 1–26 (2002).
32. F. Imamura. Review of tsunami simulation with a finite difference method. In *Long-wave runup models,* Eds. H. Yeh, C. E. Synolakis and P. L.-F. Liu (World Scientific Publishing Co., 1996), p. 25–42.
33. E. Isaacson. Water waves over a sloping bottom. *Comm. Pure Appl. Math* **3**, 11–31 (1950).
34. A. Jensen, G. Pedersen and D. J. Wood. An experimental study of wave run-up at a steep beach. *J. Fluid. Mech.* **486**, 161–188 (2003).
35. B. Johns. Numerical integration of the shallow water equations over a sloping shelf. *Int. J. Numer. Methods Fluids* **2**, 253–261 (1982).
36. H. Johnsgard. A numerical model for run-up of breaking waves. *Int. J. Num. Meth. Fluids* **31**, 1321–1331 (1999).
37. H. Johnsgard and G. Pedersen. A numerical model for three-dimensional run-up. *Int. J. Num. Meth. Fluids* **24**, 913–931 (1997).
38. U. Kânoğlu and C. E. Synolakis, Analytic solutions of solitary wave runup on the conical island and on the reverse beach. In *Long-wave runup models,* Eds. H. Yeh, C. E. Synolakis and P. L.-F. Liu (World Scientific Publishing Co., 1996), p. 215–220.
39. U. Kânoğlu and C. E. Synolakis. Long wave runup on piecewise linear topographies. *J. Fluid Mech.* **374**, 1–28 (1998).
40. T. Kataoka and M. Tsutahara. Transverse instability of surface solitary waves. *J. Fluid Mech.* **512**, 211–221 (2004).
41. H. B. Keller, A. D. Levine and G. B. Whitham. Motion of a bore over a sloping beach. *J. Fluid Mech.* **7**, 302–316 (1960).
42. A. B. Kennedy, Q. Chen, J. T. Kirby and R. A. Dalrymple. Boussinesq modeling of wave transformation, breaking, and run-up. Part I: 1D. *J. Waterw., Port, Coast., Ocean Engrg.* **126**(1), 39–47 (2000).
43. S. K. Kim, P. L.-F. Liu and J. A. Liggett. Boundary integral solutions for solitary wave generation, propagation and run-up. *Coast. Eng.* **7**, 299–317 (1983).
44. J. T. Kirby. Funwave software download page (1998).
URL: *http://chinacat.coastal.udel.edu.*
45. J. T. Kirby, G. Wei, , Q. Chen, A. B. Kennedy and R. A. Dalrymple. Fully nonlinear Boussinesq wave model documentation and user's manual. Research Report CACR-98-06, Center for applied Coastal research, Department of Civil Engineering, University of Delaware, Newark, DE 19716,

September 1998.
URL: *http://chinacat.coastal.udel.edu.*
46. N. Kobayashi, A. K. Otta and I. Roy. Wave reflection and run-up on rough slopes. *J. Waterw., Port, Coast., Ocean Engrg. Div. ASCE.* **113**, 282–298 (1987).
47. Z. Kowalik and T. S. Murty. Numerical simulation of two-dimensional tsunami run-up. *Marine Geodesy* **16**, 87–100 (1993).
48. R. J. LeVeque and D. L. George. High-resolution finite volume methods for the shallow water equations with bathymetry and dry states. In *Advanced Numerical Models for Simulating Tsunami Waves and Runup*, eds. P. L.-F. Liu, H. Yeh and C. Synolakis (World Scientific Publishing Co., 2008), p. 43–73.
49. Y. Li and F. Raichlen. Solitary wave runup on plane slopes. *J. Waterw., Port, Coast., Ocean Engrg.* **127**(1), 33–44 (2001).
50. Y. Li and F. Raichlen. Non-breaking and breaking solitary wave run-up. *J. Fluid Mech.* **456**, 295–318 (2002).
51. Pengzhi Lin, Kuang-An Chang and Philip L.-F. Liu. Runup and rundown of solitary waves on sloping beaches. *J. Waterw., Port, Coast., Ocean Engrg.* **125**(5), 247–255 (1999).
52. P. L.-F. Liu, Y.-S. Cho, M. J. Briggs, U. Kânoğlu and C. E. Synolakis. Runup of solitary waves on a circular island. *J. Fluid Mech.* **302**, 259–285 (1995).
53. P. L.-F. Liu, C. E. Synolakis and H. Yeh. Report on the international workshop on long-wave run-up. *J. Fluid Mech.* **229**, 675–688 (1991).
54. S. Longo, M. Petti and I. J. Losada. Turbulence in the swash and surf zones: a review. *Coastal Engineering* **45**, 129–147 (2002).
55. D. R. Lynch and W. G. Gray. Finite element simulation of flow in deforming regions. *J. Comp. Phys.* **36**, 135–153 (1980).
56. P. J. Lynett and P. L.-F. Liu. Coulwave model page (2004).
URL: *http://ceprofs.tamu.edu/plynett/COULWAVE/default.htm.*
57. P. J. Lynett, T.-R. Wu and P. L.-F. Liu. Modeling wave runup with depth-integrated equations. *Coast. Eng.* **46**, 89–107 (2002).
58. P. A. Madsen, H. B. Bingham and H. Liu. A new Boussinesq method for fully nonlinear waves from shallow to deep water. *J. Fluid Mech.* **462**, 1–30 (2002).
59. P. A. Madsen, R. Murray and O. R. Sörensen. A new form of the Boussinesq equations with improved linear dispersion characteristics. *Coast. Eng.* **15**, 371–388 (1991).
60. P. A. Madsen and H. A. Schäffer. Higher-order Boussinesq-type equations for surface gravity waves: derivation and analysis. *Phil. Trans. R. Soc. Lond. A* **356**, 3123–3184 (1998).
61. P. A. Madsen and H. A. Schäffer. A review of Boussinesq-type equations for surface gravity waves. *Advances in Coastal and Ocean Engineering, vol. 5* (World Scientific Publishing Co., 1999), p. 1–95.

62. P. A. Madsen, O. R. Sörensen and H. A. Schäffer. Surf zone dynamics simulated by a Boussinesq-type model: Part I. model description and and cross-shore motion of regular waves. *Coast. Eng.* **32**, 255–287 (1997).
63. J. C. Martin, W. J. Moyce, W. G. Penney, A. T. Price and C. K. Thornhill. Some gravity wave problems in the motion of perfect liquids. *Philosophical Transactions of Royal Society of London, Ser. A* **244**, 231–334 (1952).
64. F. Mesinger and A. Arakawa. Numerical methods used in atmospheric models. *GARP, Publ. Ser. WMO* **17**, (1976).
65. R. E. Meyer. On the shore singularity of water wave theory. Part 1. The local model. *Physics of Fluids* **19**, 3142–3163 (1986).
66. R. E. Meyer. On the shore singularity of water wave theory. Part 2. Small waves do not break on gentle beaches. *Physics of Fluids* **19**, 3164–3171 (1986).
67. R. E. Meyer and A. D. Taylor. Run-up on beaches. In *Waves on Beaches and Resulting Sediment Transport* (Academic Press, 1972) p. 357–411.
68. J. W. Miles. Solitary waves. *Ann. Rev. Fluid Mech.* **12**, 11–43 (1980).
69. R. L. Miller. Experimental determination of run-up of undular and fully developed bores. *J. Geophys. Res.* **73**, 4497–4510 (1968).
70. J. J. Monaghan and A. Kos. Solitary waves on a Cretan beach. *J. Waterw., Port, Coast., Ocean Engrg.* **125**(3), 145–154 (1999).
71. O. Nwogu. Alternative form of Boussinesq equations for nearshore wave propagation. *J. Waterw., Port, Coast., Ocean Engrg.* **119**(6), 618–638 (1993).
72. H. T. Özkan-Haller and J. T. Kirby. A Fourier-Chebyshev collocation method for the shallow water equations including shoreline run-up. *Applied Ocean Research* **19**, 21–34 (1997).
73. Pacific Marine Environmental Laboratory. Tsunami Reseach Program (2005),
    URL: *http://www.pmel.noaa.gov/tsunami/research.html.*
74. G. Pedersen. Run-up of periodic waves on a straight beach. Preprint Series in Applied Mathematics, ISBN 82-553-0952-7 1/85, Dept. of Mathematics, University of Oslo, Norway (1985).
75. G. Pedersen and B. Gjevik. Run-up of solitary waves. *J. Fluid Mech.* **135**, 283–299 (1983).
76. G. Pedersen. A Lagrangian model applied to runup problems. In *Advanced Numerical Models for Simulating Tsunami Waves and Runup*, eds. P. L.-F. Liu, H. Yeh and C. Synolakis (World Scientific Publishing Co., 2008), p. 311–315.
77. E. Pelinovsky, O. Kozyrev and E. Troshina. Tsunami runup in a sloping channel. In *Long-wave runup models,* Eds. H. Yeh, C. E. Synolakis and P. L.-F. Liu (World Scientific Publishing Co., 1996), p. 332–339.
78. D. H. Peregrine. Long waves on a beach. *J. Fluid Mech.* **77**, 417–431 (1967).
79. D. H. Peregrine. Equations for water waves and the approximation behind them. In *Waves on beaches*, Ed. E. Meyer (Academic Press, New York, 1972), p. 357–412.
80. D. H. Peregrine. Breaking waves on beaches. *Ann. rev. of Fluid Mech.* **15**, 149–178 (1983).

81. D. H. Peregrine and S. M. Williams. Swash overtopping a truncated plane beach. *J. Fluid Mech.* **440**, 391–399 (2001).
82. J. Petera and V. Nassehi. A new two-dimensional finite element model for the shallow water equations using a Lagrangian framework constructed along fluid particle trajectories. *Int. J. Num. Meth. in Eng.* **39**, 4159–4182 (1996).
83. M. Petti and S. Longo. Turbulence experiments in the swash zone. *Coastal Engineering* **43**, 1–24 (2001)
84. R. S. Prasad and I. A. Svendsen. Moving shoreline boundary condition for nearshore models. *Coast. Eng.* **49**(4), 239–261 (2003).
85. J. A. Puleo, R. A. Beach, R. A. Holman and J. S. Allen. Swash zone sediment suspension and transport and the importance of bore-generated turbulence. *J. Geophys. Res.* **105**(C7), 17021–17044 (2000).
86. R. O. Reid and B. R. Bodine. Numerical model for storm surges in Galveston Bay. *J. Waterw., Port, Coast., Ocean Engrg. Div. ASCE* **94**(WW1), 33–57 (1968)
87. M. C. Shen and R. E. Meyer. Climb of a bore on a beach 3, run-up. *J. Fluid Mech.* **16**, 113–125 (1963).
88. F. Shi and W. Sun. A variable boundary model of storm surge flooding in generalized curvilinear grids. *Int. J. Numer. Methods Fluids* **21**, 641–651 (1995).
89. N. Shuto. Run-up of long waves on a sloping beach. *Coastal Engng. in Japan* **10**, 23–38 (1967).
90. A. Sielecki and M. G. Wurtele. The numerical integration of the nonlinear shallow-water equations with sloping boundaries. *J. Comp. Phys.* **6**, 219–236 (1970).
91. P. A. Sleigh, P. H. Gaskell, M. Berszins and N. G. Wright. An unstructured finite-volume algorithm for predicting flow in rivers and estuaries. *Computers & Fluids* **27**(4), 479–508 (1998).
92. L. Q. Spielvogel. Single wave run-up on sloping beaches. *J. Fluid Mech.* **74**, 685–694 (1976).
93. J.J. Stoker. *Water waves.* (Interscience, 1957).
94. C. E. Synolakis. The run-up of solitary waves. *J. Fluid Mech.* **185**, 523–545 (1987).
95. C. E. Synolakis, M. K. Deb and J. E. Skjelbreia. The anomalous behaviour of the run-up of cnoidal waves. *Phys. Fluids* **31**, 3–5 (1988).
96. C. E. Synolakis and J. E. Skjelbreia. Evolution of maximum amplitude of solitary waves on plane beaches. *J. Waterw., Port, Coast., Ocean Engrg.* **119**(3), 323–342 (1993).
97. S. Tadepalli and C. Synolakis. The run-up of N-waves on sloping beaches. *Proc. R. Soc. Lond. A* **445**, 99–112 (1994).
98. T. Takagi. Finite element analysis in bench mark problems 2 and 3. In *Long-wave runup models,* Eds. H. Yeh, C. E. Synolakis and P. L.-F. Liu (World Scientific Publishing Co., 1996), p. 258–264.
99. H. Takeda. Numerical simulation of run-up by variable transformation. *J. of the Oceanographical Society of Japan* **40**, 271–278 (1984).

100. W. C. Thacker. Some exact solutions to the nonlinear shallow-water wave equations. *J. Fluid Mech.* **107**, 499–508 (1981).
101. V. V. Titov and C. E. Synolakis. Modeling of breaking and nonbreaking long-wave evolution and runup using VTCS-2. *J. Waterw., Port, Coast., Ocean Engrg.* **121**(6), 308–316 (1995).
102. V. V. Titov and C. E. Synolakis. Numerical modeling of tidal wave runup. *J. Waterw., Port, Coast., Ocean Engrg.* **124**(4), 157–171 (1998).
103. E. O. Tuck and L.-S. Hwang. Long wave generation on a sloping beach. *J. Fluid Mech.* **51**(3), 449–461 (1972).
104. T. Umetsu. A boundary condition technique of moving boundary simulation for broken dam problem by three-step explicit finite element method. In *Advances in Hydro-Science and Engineering, Vol. II*, (Tsinghua University Press, Beijing, 1995), p. 394–399.
105. A. Walters and T. Takagi. Review of finite element methods for tsunami simulations. In *Long-wave runup models*, Eds. H. Yeh, C. E. Synolakis and P. L.-F. Liu (World Scientific Publishing Co., 1996), p. 43–87.
106. G. Watson, D. H. Peregrine and E. F. Toro. Numerical solution of the shallow water equations on a beach using the weighted average flux method. In *Computational Fluid Dynamics '92* Eds. Ch. Hirsch, J. Pèriaux and W. Kordulla (Elsevier Science Publishers, 1992), p. 495–502.
107. G. Wei, J. T. Kirby, S. T. Grilli and R. Subramanya. A fully nonlinear Boussinesq model for surface waves. Part 1. Highly nonlinear unsteady waves. *J. Fluid Mech.* **294**, 71–92 (1995).
108. J.-K. Woo and P. L.-F. Liu. Finite element model for modified Boussinesq equations. I: Model development. *J. Waterw., Port, Coast., Ocean Engrg.* **130**(1), 1–16 (2004).
109. D.J. Wood, G. Pedersen and A. Jensen. Modelling of run-up of steep nonbreaking waves. *Ocean Engineering* **30**, 625–644 (2003).
110. T. Y. Wu. Long waves in ocean and coastal waters. *Proc. ASCE, J. Eng. Mech. Div.* **107**, 501–522 (1981).
111. H. Yeh, A. Ghazali and I. Marton. Experimental study of a bore run-up. *J. Fluid Mech.* **206**, 563–578 (1989).
112. H. Yeh, P. L.-F. Liu, M. Briggs and C. E. Synolakis. Propagation and amplification of tsunamis at coastal boundaries. *Nature* **372**, 353–355 (1994).
113. H. Yeh, C. E. Synolakis and P. L.-F. Liu, Eds., *Long-wave runup models.* (World Scientific Publishing Co., 1996).
114. J. A. Zelt. The run-up of nonbreaking and breaking solitary waves. *Coast. Eng.* **15**, 205–246 (1991).
115. J. A. Zelt and F. Raichlen. A Lagrangian model for wave-induced harbour oscillations. *J. Fluid Mech.* **213**, 203–225 (1990).
116. J. A. Zelt and F. Raichlen. Overland flow from solitary waves. *J. Waterw., Port, Coast., Ocean Engrg.* **117**, 247–263 (1991).
117. X. Zhou, Y. C. Hon and K. F. Cheung. A grid-free, nonlinear shallow-water model with moving boundary. *Engineering Analysis with Boundary Elements* **28**(8), 967–973 (2004).

CHAPTER 2

HIGH-RESOLUTION FINITE VOLUME METHODS FOR THE SHALLOW WATER EQUATIONS WITH BATHYMETRY AND DRY STATES

Randall J. LeVeque[1] and David L. George[2]

Department of Applied Mathematics, University of Washington
Box 352420, Seattle, WA 98195-2420
E-mail: [1]*rjl@amath.washington.edu,* [2]*dgeorge@amath.washington.edu*

We give a brief review of the wave-propagation algorithm, a high-resolution finite volume method for solving hyperbolic systems of conservation laws. These methods require a Riemann solver to resolve the jump in variables at each cell interface into waves. We present a Riemann solver for the shallow water equations that works robustly with bathymetry and dry states. This method is implemented in CLAWPACK and applied to benchmark problems from the Third International Workshop on Long-Wave Runup Models, including a two-dimensional simulation of runup during the 1993 tsunami event on Okushiri Island. Comparison is made with wave tank experimental data provided for the workshop. Some preliminary results using adaptive mesh refinement on the 26 December 2004 Sumatra event are also presented.

1. Introduction

We will present a brief introduction to a class of high-resolution finite volume methods for hyperbolic problems and discuss the application of these methods to long-wave runup problems using the shallow water equations. To solve the benchmark problems for this workshop we have used such methods in one and two space dimensions that work robustly with bathymetry (bottom topography) and dry states, and that automatically handle the moving interface between water and land. Some results on the benchmark problems are presented in Secs. 6 and 8, and more results, along with some animations, may be found at the website[15].

We use a mathematical framework known as the wave-propagation algorithm that has been implemented in the software package CLAWPACK (Conservation Laws Package) in 1, 2, and 3 space dimensions, and which also includes adaptive mesh refinement capabilities. This algorithm gives a general formulation of a class of finite volume methods known as "high-resolution shock-capturing Godunov-type methods" that are second order accurate in space and time on smooth solutions while automatically capturing discontinuities in the solution (including shocks or hydraulic jumps) with minimal numerical smearing and no spurious oscillations. More details on these algorithms and their application to hyperbolic problems may be found in LeVeque[12,14] and the CLAWPACK software documentation, available at www.clawpack.org.

The fact that these are finite volume methods means that, rather than pointwise approximations to the solution, the numerical solution consists of approximations to the *cell averages* of the solution over grid cells. Here the grid cells are assumed to be intervals $[x_{i-1/2}, x_{i+1/2}]$ of uniform length Δx in one dimension or rectangles $[x_{i-1/2}, x_{i+1/2}] \times [y_{j-1/2}, y_{j+1/2}]$ in two dimensions, but more general nonuniform grids can also be used.

Finite volume methods are particularly appropriate when solving systems of conservation laws, in which case the integral of the solution over each grid cell is modified only due to fluxes through the edges of the grid cell. Dividing this statement by the cell area leads to an update formula for the cell averages based on numerical approximations to the flux through each edge, as written out below. Such a finite volume method is based directly on the integral form of the conservation law and can be applied to problems with discontinuous solutions more reliably than finite difference approximations to the differential equation form of the conservation law, which does not hold at a discontinuity.

In one space dimension the integral form of a conservation law is

$$\frac{d}{dt} \int_{x_1}^{x_2} q(x,t)\, dx = f(q(x_1,t)) - f(q(x_2,t)) \qquad \forall x_1, x_2, \tag{1}$$

where $q(x,t) \in \mathbb{R}^m$ is the vector of conserved quantities and $f(q)$ is the flux function. This states that the total mass of q in any interval $[x_1, x_2]$ changes only due to fluxes through the edges.

The shallow water equations on a flat bottom have this form with $m = 2$ and

$$q = \begin{bmatrix} h \\ hu \end{bmatrix}, \qquad f(q) = \begin{bmatrix} hu \\ hu^2 + \frac{1}{2}gh^2 \end{bmatrix}, \tag{2}$$

where h is the fluid depth, u is the horizontal velocity, and g is the gravitational constant. These equations express the conservation of mass and momentum.

If the solution is sufficiently smooth, then the integral conservation law (1) can be manipulated to yield

$$\int_{x_1}^{x_2} q_t(x,t) + f(q(x,t))_x = 0 \quad \forall x_1, x_2,$$

and hence

$$q_t + f(q)_x = 0, \tag{3}$$

which is the PDE form of the conservation law (with subscripts denoting partial derivatives). This equation is called *hyperbolic* if the Jacobian matrix $f'(q_0) \in \mathbb{R}^{m \times m}$ is diagonalizable and has real eigenvalues for any physically relevant state q_0. Hyperbolic problems typically model wave propagation and the eigenvalues correspond to the propagation velocities if we linearize about the state q_0. For the shallow water equations given by (2),

$$f'(q) = \begin{bmatrix} 0 & 1 \\ -u^2 + gh & 2u \end{bmatrix}, \tag{4}$$

with eigenvalues

$$\lambda^1 = u - \sqrt{gh}, \quad \lambda^2 = u + \sqrt{gh}, \tag{5}$$

and corresponding eigenvectors

$$r^1 = \begin{bmatrix} 1 \\ u - \sqrt{gh} \end{bmatrix} = \begin{bmatrix} 1 \\ \lambda^1 \end{bmatrix}, \quad r^2 = \begin{bmatrix} 1 \\ u + \sqrt{gh} \end{bmatrix} = \begin{bmatrix} 1 \\ \lambda^2 \end{bmatrix}. \tag{6}$$

A finite volume method in *conservation form* updates the cell average Q_i^n of the solution over the grid cell using an expression

$$Q_i^{n+1} = Q_i^n - \frac{\Delta t}{\Delta x}[F_{i+1/2}^n - F_{i-1/2}^n], \tag{7}$$

where

$$Q_i^n \approx \frac{1}{\Delta x} \int_{x_{i-1/2}}^{x_{i+1/2}} q(x, t_n)\, dx,$$

$$F_{i-1/2}^n \approx \frac{1}{\Delta t} \int_{t_n}^{t_{n+1}} f(q(x_{i-1/2}, t))\, dt \tag{8}$$

are the numerical approximations to the cell average and interface flux, respectively. The update (7) comes directly from integrating (1) in time from t_n to t_{n+1} and dividing by Δx. Equation (7) can also be viewed as a direct discretization of the PDE (3), but viewing the value $F_{i-1/2}^n$ as an approximation to the interface flux is key in developing high-resolution methods for nonlinear problems.

We will generally drop the superscript n on quantities at time t_n to simplify the notation below, particularly since other superscripts will be needed.

1.1. *Godunov's Method*

The methods we use are extensions of Godunov's method, a first-order method for gas dynamics developed in the 1950s in which the interface flux $F_{i-1/2}$ is computed by solving a *Riemann problem* between the states Q_{i-1} and Q_i. This is simply the conservation law with piecewise constant data (as in a shock tube or dam break problem). Such a problem naturally arises at the cell interface if the solution at time t_n is approximated by a piecewise constant function with values Q_j at all points in the jth grid cell.

Godunov's method is applicable to any hyperbolic conservation law. For a linear problem in which $f(q) = Aq$ for some matrix A (so $f'(q) = A$), the solution to the Riemann problem for any states Q_{i-1} and Q_i consists of m discontinuities (waves), each proportional to an eigenvector r^p of A, and propagating at speed equal to the corresponding eigenvalue λ^p (for $p = 1, 2, \ldots, m$). Since A must be diagonalizable, we can write

$$A = R\Lambda R^{-1},$$

where $\Lambda = \text{diag}(\lambda^1, \ldots, \lambda^m)$ and $R = [r^1 \cdots r^m]$ is the invertible matrix of eigenvectors.

Solving the Riemann problem between states Q_{i-1} and Q_i is then easily accomplished by decomposing $\Delta Q_{i-1/2} = Q_i - Q_{i-1}$ into eigenvectors of A, i.e., writing $\Delta Q_{i-1/2}$ as a linear combination of the vectors r^p,

$$Q_i - Q_{i-1} = \sum_{p=1}^{m} \alpha^p_{i-1/2} r^p \equiv \sum_{p=1}^{m} \mathcal{W}^p_{i-1/2}. \tag{9}$$

We use $\mathcal{W}^p$ to denote the pth wave in this Riemann solution. The vector of coefficients $\alpha^p_{i-1/2}$ is given by $\alpha_{i-1/2} = R^{-1}\Delta Q_{i-1/2}$. Godunov's method in the linear case is then defined by setting

$$F_{i-1/2} = f(Q^{\downarrow}_{i-1/2}) = AQ^{\downarrow}_{i-1/2},$$

where $Q^{\downarrow}_{i-1/2}$ denotes the value at the interface $x_{i-1/2}$ in the Riemann solution,

$$Q^{\downarrow}_{i-1/2} = Q_{i-1} + \sum_{p:\lambda^p<0} \mathcal{W}^p_{i-1/2}.$$

Multiplying by A gives

$$F_{i-1/2} = AQ_{i-1} + \sum_{p=1}^{m} \alpha_{i-1/2}^{p} (\lambda^p)^- r^p \qquad (10)$$
$$= AQ_{i-1} + A^- \Delta Q_{i-1/2},$$

where

$$A^- = R\,\text{diag}((\lambda^p)^-)\,R^{-1}, \qquad \text{with} \;\; \lambda^- = \min(\lambda, 0). \qquad (11)$$

Alternatively, we can write

$$Q_{i-1/2}^{\downarrow} = Q_i - \sum_{p:\lambda^p > 0} \mathcal{W}_{i-1/2}^p$$

and obtain

$$F_{i-1/2} = AQ_i + \sum_{p=1}^{m} \alpha_{i-1/2}^{p} (\lambda^p)^+ r^p \qquad (12)$$
$$= AQ_i + A^+ \Delta Q_{i-1/2},$$

where

$$A^+ = R\,\text{diag}((\lambda^p)^+)\,R^{-1}, \qquad \text{with} \;\; \lambda^+ = \max(\lambda, 0). \qquad (13)$$

For a linear problem the resulting method is simply the first-order upwind method, extended from the scalar advection equation to a general system by diagonalizing the system and applying the upwind method to each characteristic component in the appropriate direction based on the propagation velocity.

For nonlinear problems, such as the shallow water equations, the exact solution to the Riemann problem is harder to calculate but can still be worked out (see, e.g. LeVeque[14] and Toro[19]) and the resulting interface flux used for $F_{i-1/2}$. In practice, however, it is usually more efficient to use some *approximate Riemann solver* to obtain $F_{i-1/2}$. One popular choice is to use a "Roe solver" following the work of Roe[17] for gas dynamics, in which the data Q_{i-1}, Q_i is used to define a "Roe-averaged" Jacobian matrix $A_{i-1/2}$ by a suitable combination of $f'(Q_{i-1})$ and $f'(Q_i)$. The numerical flux is then determined by solving the Riemann problem for the linear problem $q_t + A_{i-1/2} q_x = 0$. The Roe average is chosen to have the property that

$$f(Q_i) - f(Q_{i-1}) = A_{i-1/2}(Q_i - Q_{i-1}). \qquad (14)$$

This leads to nice properties in the approximate solution. The Roe matrix for the shallow water equations is easily computed (see, e.g. LeVeque[14]), and is simply the Jacobian matrix (4) evaluated at the Roe-averaged state

$$\hat{h}_{i-1/2} = \frac{h_i + h_{i-1}}{2}, \qquad \hat{u}_{i-1/2} = \frac{u_{i-1}\sqrt{gh_{i-1}} + u_i\sqrt{gh_i}}{\sqrt{gh_{i-1}} + \sqrt{gh_i}}. \qquad (15)$$

The eigenvalues, or "Roe speeds", are therefore

$$\hat{s}^1_{i-1/2} = \hat{u}_{i-1/2} - \sqrt{g\hat{h}_{i-1/2}}, \qquad \hat{s}^2_{i-1/2} = \hat{u}_{i-1/2} + \sqrt{g\hat{h}_{i-1/2}}, \qquad (16)$$

and the Roe eigenvectors are

$$\hat{r}^1_{i-1/2} = \begin{bmatrix} 1 \\ \hat{s}^1_{i-1/2} \end{bmatrix}, \qquad \hat{r}^2_{i-1/2} = \begin{bmatrix} 1 \\ \hat{s}^2_{i-1/2} \end{bmatrix}. \qquad (17)$$

In one dimension the CLAWPACK software requires a Riemann solver that, for any states Q_{i-1} and Q_i, returns a set of M_w waves $\mathcal{W}^1_{i-1/2}, \ldots, \mathcal{W}^{M_w}_{i-1/2}$, and propagation speeds for the waves, $s^1_{i-1/2}, \ldots, s^{M_w}_{i-1/2}$. The number of waves M_w may be equal to m, the dimension of the system, but could be different. The Riemann solver must also return the *fluctuations* $\mathcal{A}^-\Delta Q_{i-1/2}$ and $\mathcal{A}^+\Delta Q_{i-1/2}$, two vectors that are used to update the solution according to

$$Q_i^{n+1} = Q_i - \frac{\Delta t}{\Delta x}(\mathcal{A}^+\Delta Q_{i-1/2} + \mathcal{A}^-\Delta Q_{i+1/2}). \qquad (18)$$

These fluctuations should have the property that

$$\mathcal{A}^-\Delta Q_{i-1/2} + \mathcal{A}^+\Delta Q_{i-1/2} = f(Q_i) - f(Q_{i-1}), \qquad (19)$$

so that they represent a "flux difference splitting". The fluctuations may be defined in terms of the interface fluxes as

$$\begin{aligned}\mathcal{A}^+\Delta Q_{i-1/2} &= f(Q_i) - F_{i-1/2}, \\ \mathcal{A}^-\Delta Q_{i+1/2} &= F_{i+1/2} - f(Q_i).\end{aligned} \qquad (20)$$

Then (19) is satisfied (note the shift in index) and (18) reduces to the flux-differencing update formula (7). The form (18) is used in CLAWPACK and the general formulation of the wave-propagation algorithms because it is more flexible and allows the extension of these methods to hyperbolic problems that are not in conservation form, in which case there is no "flux function" (see LeVeque[14]).

The notation $\mathcal{A}^{\pm}\Delta Q_{i-1/2}$ used for the fluctuation vectors is suggested by the fact that, for a linear problem, the natural choice is

$$\mathcal{A}^{-}\Delta Q_{i-1/2} = A^{-}(Q_i - Q_{i-1}),$$
$$\mathcal{A}^{+}\Delta Q_{i-1/2} = A^{+}(Q_i - Q_{i-1}), \tag{21}$$

where the matrices $A^{\pm}$ are defined by (13) and (11). In general they are often defined by

$$\mathcal{A}^{-}\Delta Q_{i-1/2} = \sum_{p=1}^{M_w} (s_{i-1/2}^p)^{-} \mathcal{W}_{i-1/2}^p,$$
$$\mathcal{A}^{+}\Delta Q_{i-1/2} = \sum_{p=1}^{M_w} (s_{i-1/2}^p)^{+} \mathcal{W}_{i-1/2}^p. \tag{22}$$

For a linear problem or a nonlinear problem when the Roe solver is used, this agrees with the definition (20).

The first-order method (18) only uses the fluctuations returned from the Riemann solver, and does not make explicit use of the waves $\mathcal{W}^p$ or speeds s^p. These quantities are used in the high-resolution correction terms discussed in the next section.

1.2. *High-Resolution Corrections*

The method (18) is only first-order accurate. The second-order Lax-Wendroff method for the linear problem can be written as a modification to (18), as

$$Q_i^{n+1} = Q_i - \frac{\Delta t}{\Delta x}(\mathcal{A}^{+}\Delta Q_{i-1/2} + \mathcal{A}^{-}\Delta Q_{i+1/2}) - \frac{\Delta t}{\Delta x}(\tilde{F}_{i+1/2} - \tilde{F}_{i-1/2}), \tag{23}$$

where the fluctuations are still given by (21) for the linear problem and the correction fluxes are

$$\tilde{F}_{i-1/2} = \frac{1}{2}\left(|A| - \frac{\Delta t}{\Delta x}A^2\right)\Delta Q_{i-1/2} = \frac{1}{2}\left(I - \frac{\Delta t}{\Delta x}|A|\right)|A|\Delta Q_{i-1/2}, \tag{24}$$

where $|A| = A^{+} - A^{-}$. Inserting this in (23) and manipulating yields a more familiar form of the Lax-Wendroff method,

$$Q_i^{n+1} = Q_i - \frac{\Delta t}{2\Delta x}A(Q_{i+1} - Q_{i-1}) + \frac{\Delta t^2}{2\Delta x^2}A^2(Q_{i+1} - 2Q_i + Q_{i-1}). \tag{25}$$

This method is second-order accurate on smooth solutions but is highly dispersive and nonphysical oscillations arise near steep gradients or discontinuities in the solution, which can completely destroy the accuracy. These oscillations can be avoided by using the form (23) and applying appropriate limiters to the correction terms. To this end we rewrite (24) as

$$\tilde{F}_{i-1/2} = \frac{1}{2} \sum_{p=1}^{m} \left(I - \frac{\Delta t}{\Delta x} |\lambda^p| \right) |\lambda^p| \widetilde{\mathcal{W}}^p_{i-1/2}, \qquad (26)$$

where $\mathcal{W}^p_{i-1/2} = \alpha^p_{i-1/2} r^p$ are the waves obtained from the Riemann solution, as in (9), and $\widetilde{\mathcal{W}}^p_{i-1/2}$ represents a "limited" version of the wave. Each wave is limited by comparing it to the wave in the same family arising from the Riemann problem at the neighboring interface in the upwind direction, i.e. we set

$$\widetilde{\mathcal{W}}^p_{i-1/2} = \text{limiter}(\mathcal{W}^p_{i-1/2}, \mathcal{W}^p_{I-1/2}), \qquad (27)$$

where

$$I = \begin{cases} i-1 & \text{if } \lambda^p > 0 \\ i+1 & \text{if } \lambda^p < 0. \end{cases}$$

If $\mathcal{W}^p_{I-1/2}$ and $\mathcal{W}^p_{i-1/2}$ are "comparable" in some sense then this component of the solution is presumed to be smoothly varying. In this case the corresponding term in (26) can be expected to give a useful correction that will help improve accuracy and the limiter should return $\widetilde{\mathcal{W}}^p_{i-1/2} \approx \mathcal{W}^p_{i-1/2}$. However, if $\mathcal{W}^p_{I-1/2}$ and $\mathcal{W}^p_{i-1/2}$ quite different then this component is not smoothly varying and attempting to add an additional term from the Taylor series may make things worse rather than better. In this case the limiter should modify the wave, typically be reducing its magnitude. There is an extensive theory of limiters that will not be further discussed here.

The method (23) with correction fluxes (26) is easily extended to nonlinear problems. Recall that the (approximate) Riemann solver returns fluctuations $\mathcal{A}^{\pm} \Delta Q_{i-1/2}$, waves $\mathcal{W}^p_{i-1/2}$, and speeds $s^p_{i-1/2}$. The only change in the formulas required in order to apply (23) to a general nonlinear problem is to replace λ^p in (26) by the local speed $s^p_{i-1/2}$, obtaining

$$\tilde{F}_{i-1/2} = \frac{1}{2} \sum_{p=1}^{M_w} \left(I - \frac{\Delta t}{\Delta x} |s^p_{i-1/2}| \right) |s^p_{i-1/2}| \widetilde{\mathcal{W}}^p_{i-1/2}. \qquad (28)$$

2. The f-Wave Approach

The method described above can be reformulated in a way that will prove particularly useful when source terms are added to the equations, as needed to handle variable bathymetry. Recall that the waves $\mathcal{W}^p_{i-1/2}$ correspond to a splitting of the jump in Q at the interface $x_{i-1/2}$,

$$Q_i - Q_{i-1} = \sum_{p=1}^{M_w} \mathcal{W}^p_{i-1/2}.$$

In a linear problem, or a nonlinear problem that has been locally linearized using a Roe matrix $A_{i-1/2}$, these waves are obtained by expressing $\Delta Q_{i-1/2}$ as a linear combination of the eigenvectors $r^p_{i-1/2}$ of the matrix, i.e., $\mathcal{W}^p_{i-1/2} = \alpha^p_{i-1/2} r^p_{i-1/2}$ for some scalars $\alpha^p_{i-1/2}$, as is done in (9). Alternatively, we could split the jump in $f(Q)$ into eigencomponents as

$$f(Q_i) - f(Q_{i-1}) = \sum_{p=1}^{M_w} \beta^p_{i-1/2} r^p_{i-1/2} \equiv \sum_{p=1}^{M_w} \mathcal{Z}^p_{i-1/2}.$$

If the matrix $A_{i-1/2}$ satisfies Roe's condition (14), then we simply have $\mathcal{Z}^p_{i-1/2} = s^p_{i-1/2} \mathcal{W}^p_{i-1/2}$. For other approximate Riemann solvers it is necessary to determine an appropriate splitting of $f(Q)$ based on the splitting of Q.

The vectors $\mathcal{Z}^p$ are called f-waves because they carry jumps in f rather than jumps in q. Since $\mathcal{Z}^p_{i-1/2} = \text{sgn}(s^p_{i-1/2})|s^p_{i-1/2}|\mathcal{W}^p_{i-1/2}$ for linearized Riemann solvers, the natural generalization of the correction flux (28) for the f-wave formulation is

$$\tilde{F}_{i-1/2} = \frac{1}{2} \sum_{p=1}^{M_w} \left(I - \frac{\Delta t}{\Delta x} \text{sgn}(s^p_{i-1/2}) \right) \widetilde{\mathcal{Z}}^p_{i-1/2}, \qquad (29)$$

where $\widetilde{\mathcal{Z}}^p_{i-1/2}$ is a limited version of $\mathcal{Z}^p_{i-1/2}$ calculated using the same limiter as previously applied to $\mathcal{W}^p$.

One advantage of the f-wave approach is that any linearly independent set of vectors r^p_{i-1} can be used to define the splitting of Δf into waves $\mathcal{Z}^p_{i-1/2}$ and the method remains conservative. Of course a reasonable choice is required in order to maintain consistency with the differential equation, but for example the eigenvectors of any reasonable approximate Jacobian matrix based on Q_{i-1} and Q_i could be used without needing to impose the Roe condition (14). For the shallow water equations this suggests using vectors

$$r^1 = \begin{bmatrix} 1 \\ s^1 \end{bmatrix}, \qquad r^2 = \begin{bmatrix} 1 \\ s^2 \end{bmatrix}, \qquad (30)$$

where s^1 and s^2 are some approximations to the wave speeds of the two waves in the Riemann solution. Taking s^1 and s^2 to be the Roe speeds

(16) recovers the Roe solver, but in some cases this is not a good choice. In particular the Roe solver can fail when dry states are expected in the solution. In Sec. 4 we present a different choice of speeds that can be used much more robustly.

3. Bathymetry and Source Terms

We now consider the shallow water equations over non-constant bathymetry with elevation $y = B(x)$. In this case $h(x,t)$ represents the depth of the water above the bathymetry. A sloping bottom can accelerate the fluid and gives rise to a source term in the momentum equation proportional to the slope $B_x(x)$. The shallow water equations now take the form

$$q_t + f(q)_x = \psi(q, x), \tag{31}$$

where q and f are as in (2) and the source term is

$$\psi(q, x) = \begin{bmatrix} 0 \\ -ghB_x \end{bmatrix}. \tag{32}$$

One way to tackle (31) numerically is is to use a *fractional step* procedure. In each time step one first solves the homogeneous conservation law (3) to advance Q^n by Δt to obtain an intermediate state Q^*, and then solves

$$q_t = \psi(q, x) \tag{33}$$

over time Δt to advance Q^* to Q^{n+1}. This often works well, but is subject to numerical difficulties, particularly in situations where there is a steady state with $f(q) = \psi(q, x)$ and the desired dynamic solution is a relatively small perturbation of this steady state. In this case solving (3) and (33) may each lead to significant changes in the solution that should nearly cancel out. Numerically, the use of distinct numerical methods in separate steps can lead to errors that swamp the desired solution when the waves of interest are small perturbations of the steady state.

Instead of using a fractional step method, we modify the f-wave formulation of the hyperbolic solver and incorporate the source term into the flux difference before decomposing this into waves, i.e. we decompose

$$f(Q_i) - f(Q_{i-1}) - \Delta x \Psi_{i-1/2} = \sum_{p=1}^{M_w} \mathcal{Z}_{i-1/2}^p, \tag{34}$$

where $\Psi_{i-1/2}$ is a discretization of the source term. For the shallow water equations with bathymetry the source term $\Delta x g h B_x$ at $x_{i-1/2}$

is approximated by $\frac{1}{2}g(h_{i-1} + h_i)(B_i - B_{i-1})$, resulting in the vector $f(Q_i) - f(Q_{i-1}) - \Delta x \Psi_{i-1/2}$ taking the form

$$\begin{bmatrix} h_i u_i - h_{i-1} u_{i-1} \\ \left(h_i u_i^2 + \frac{1}{2}gh_i^2\right) - \left(h_{i-1} u_{i-1}^2 + \frac{1}{2}gh_{i-1}^2\right) + \frac{1}{2}g(h_{i-1} + h_i)(B_i - B_{i-1}) \end{bmatrix}. \tag{35}$$

This vector is decomposed into f-waves, for example by writing it as a linear combination of the eigenvectors $r_{i-1/2}^p, (p = 1, 2)$ of the Roe matrix or of (30).

Of particular importance is the special case of a motionless body of water over variable bathymetry, in which case $u \equiv 0$ and $h(x,t) + B(x)$ is initially constant and should remain so. In this case the vector (35) becomes

$$\begin{bmatrix} 0 \\ \frac{1}{2}g\left(h_i^2 - h_{i-1}^2\right) + \frac{1}{2}g(h_{i-1} + h_i)(B_i - B_{i-1}) \end{bmatrix}. \tag{36}$$

If $h_j + B_j$ is constant in j then $B_i - B_{i-1} = h_{i-1} - h_i$ and (36) is the zero vector, resulting in $\mathcal{Z}_{i-1/2}^1 = \mathcal{Z}_{i-1/2}^2 = 0$ and $\mathcal{A}^{\pm}\Delta Q_{i-1/2} = 0$ in (34). Hence $Q_i^{n+1} = Q_i$ and the steady state is exactly maintained numerically.

Moreover, if we are modeling waves in which perturbations in $h + B$ are small compared to variations in B, then this vector captures the perturbations after canceling out the steady state portion of the variation. The wave limiters and high resolution correction terms are applied to these perturbations rather than to waves that include the large variation in h due to the bathymetry. As a result, this method is much more accurate in modeling the propagation of small amplitude perturbations than a fractional step method. This may be quite important in calculating the long-range propagation of small-amplitude tsunamis through the ocean.

The f-wave approach and its use for source terms is discussed further in Bale et al.[1] and LeVeque[14]. A variety of other approaches have also been developed recently for this problem, for example Gosse[7], Greenberg et al.[8], Jenny and Müller[10], Kurganov and Levy[11], LeVeque[13].

4. Dry States

It is well known that if the Roe solver is used to solve a Riemann problem in which $u_{i-1} < u_i$ with sufficient difference in velocities, then the approximate Riemann solution will have a negative depth in the intermediate state (see Fig. 15.3 in LeVeque[14] for an illustration of this). This nonphysical behavior often causes the computation to crash. In reality a dry state should form if the velocity difference is sufficiently large, although the Roe solver can fail even for smaller velocity differences (see Einfeldt et al.[6] for some discussion of related issues for the Euler equations). Similar problems arise when solving a Riemann problem with one state initially dry, $h_{i-1} = 0$ or $h_i = 0$, as happens at some cell interfaces for any problem involving wave motion at the shore.

This difficulty can be avoided by using the f-wave approach discussed in Secs. 2 and 3 using eigenvectors r^1 and r^2 from (30) with a better choice of speeds than the Roe speeds. In most cases we use the "Einfeldt speeds"

$$s_E^1 = \min(u_{i-1} - \sqrt{gh_{i-1}}, \hat{s}^1), \qquad s_E^2 = \max(u_i + \sqrt{gh_i}, \hat{s}^2), \qquad (37)$$

obtained by comparing the Roe speeds with the characteristic speeds λ_{i-1}^1 and λ_i^1 (the eigenvalues of the Jacobian matrices in states Q_{i-1} and Q_i). This choice is adapted from the suggestion of Einfeldt[5] that speeds corresponding to these be used in the HLL method for gas dynamics. The HLL approximate Riemann solver (after Harten et al.[9]) simply uses two waves to approximate the Riemann solution (regardless of the dimension m of the system) with speeds s^1 and s^2 approximating the minimum and maximum wave speeds arising in the system. The HLLE method, using the Einfeldt speeds, chooses these speeds by comparing the Roe speed, a reasonable choice if the wave is a shock, and the extreme characteristic speed, which may be faster if the wave is instead a spreading rarefaction wave. Since $m = 2$ in the one-dimensional shallow water equations, the method we use is closely related to the HLLE method, though not the same and will be modified further to handle source terms and dry states below. (See LeVeque and Pelanti[16] for some more discussion of the relation between the HLL and Roe solvers and the f-wave approach.)

Using the f-wave approach with the choice of speeds (37) and corresponding eigenvectors (30) nearly always maintains non-negative depth. In fact, it can be shown that the total mass in the intermediate Riemann solution is always positive given these speeds and eigenvectors. However, as explained below, it is sometimes necessary to further modify the wave speeds to maintain non-negativity, at least in the subcritical case when $s_E^1 < 0 < s_E^2$. In the supercritical case when both speeds have the same sign, no modification is needed.

We first consider the case of preserving non-negativity in a cell that has positive depth initially, $h_{i-1} > 0$ or $h_i > 0$, and later consider the case of preserving non-negativity in a cell that is already initially dry. Even in the case where both cells have a positive depth initially, $h_{i-1} > 0$ and $h_i > 0$, a negative depth can be generated if $s_E^1 < 0 < s_E^2$ when the choice (37) is used. Although the total mass in the intermediate Riemann solution is positive, it may happen that the approximate Riemann solution leads to the mass going negative on one side of the interface. In this case it can be shown that the mass must be increasing on the other side by at least the same amount, and so negativity can be avoided by a transfer of mass that can be accomplished by increasing the speed on the side losing mass. Working out the formula to increase this speed just to the point where the negative state reaches $h = 0$, we find that in general the following speeds can always be used:

$$s^1 = \min\left(s_E^1, \; s_E^2/2 - \sqrt{(s_E^2/2)^2 + \max(0, s_E^2 \Delta^1 - \Delta^2)/h_{i-1}}\right), \quad (38)$$

$$s^2 = \max\left(s_E^2, \; s_E^1/2 + \sqrt{(|s_E^1|/2)^2 + \max(0, \Delta^2 - s_E^1 \Delta^1)/h_i}\right). \quad (39)$$

That is, given $h_{i-1} > 0$ initially, (38) will maintain that $h_{i-1} \geq 0$, and given $h_i > 0$ initially (39) will maintain that $h_i \geq 0$. Here Δ^1 and Δ^2 are the components of the modified flux difference

$$f(Q_i) - f(Q_{i-1}) - \Delta x \Psi_{i-\frac{1}{2}}, \quad (40)$$

which takes into account the bathymetry. Each speed (38) and (39) always corresponds to the Einfeldt speed unless a negative state would arise on that side, so for a given Riemann problem, at most one of s^1 and s^2 is different from the Einfeldt speed, and only when necessary to maintain non-negativity.

Maintaining the non-negativity in a cell with a depth that is initially 0 is handled somewhat differently. If $B_{i-1} = B_i$, and only one of h_{i-1} or h_i is positive, then the true solution to this Riemann problem consists only of a rarefaction wave with the leading edge propagating at speed $u_i - 2\sqrt{gh_i}$ if $h_{i-1} = 0$ or speed $u_{i-1} + 2\sqrt{gh_{i-1}}$ if $h_i = 0$. We use these speeds as s^1 or s^2 if $h_{i-1} = 0$ or $h_i = 0$ respectively. As stated above, using the speed (38) will preserve $h_{i-1} \geq 0$ if it is initially positive, and (39) will preserve $h_i \geq 0$ if it is initially positive. Using the speed of the leading edge of a rarefaction wave for the complimentary speed however, will not necessarily prevent the dry state from becoming negative. It can be shown however that negativity is only possible if the true rarefaction wave is transonic ($s^1 < 0 < u_{i-1} + 2\sqrt{gh_{i-1}}$ if $h_i = 0$ or $u_i - 2\sqrt{gh_i} < 0 < s^2$ if $h_{i-1} = 0$). Transonic rarefactions are a more general problem, and are discussed in the following section.

5. Entropy Conditions

Integral conservation laws can be satisfied by discontinuous weak solutions subject to the Rankine-Hugoniot jump conditions. This results in possible non-uniqueness of weak solutions — an initial value problem might be satisfied by both a smooth solution and a discontinuous one. Determining the physically relevant solution requires additional admissibility conditions, often taking the form of an "entropy" function that is conserved except across a discontinuity. The name arises from the Euler equations of gas dynamics, where the entropy must increase when gas passes through a shock. For the shallow water equations, the "entropy" function is actually mechanical energy, which must decrease when passing through a discontinuity.

Since the integral conservation laws alone do not guarantee a unique solution to an initial value problem, a numerical method based on the conservation laws alone might converge to an entropy violating weak solution. An "entropy fix" is therefore often needed. Determining such a fix for Godunov-type methods requires a careful look at the particular Riemann solver being used. If an approximate solver is used, true solutions to Riemann problems which consist of m waves — any combination of rarefactions and shock waves, are replaced by m jump discontinuities locally at each grid cell. These discontinuities might approximate a physically relevant shock wave, or perhaps a smooth rarefaction. In the latter case, the jump discontinuity still approximates the conservation law, however it more closely resembles the entropy violating shock than the physically relevant rarefaction.

With the wave propagation methods, the waves arising from a particular Riemann problem at a grid cell interface are averaged onto the adjacent cells. Therefore, the local discrepancy between an entropy violating discontinuity and a smooth rarefaction may have no effect on the numerical solution, if both produce the same average within a grid cell. This will be the case if the wave structure of the rarefaction remains entirely within a grid cell. The exceptional case is a transonic rarefaction — a rarefaction where one of the eigenvalues passes through zero. This type of rarefaction has a wave structure that should overlap a cell interface, yet it is approximated by a jump discontinuity moving either to the left or the right. This does affect the numerical solution, and can cause a method to converge globally to an entropy violating weak solution. The fix is to determine when the correct entropy solution to a Riemann problem corresponds to a transonic rarefaction, and then split the entropy violating single wave, apportioning it to the adjacent grid cells appropriately.

An alternative and numerically more useful formulation of the entropy condition for the shallow water equations is that a physically correct shock in the pth ($p = 1, 2$) characteristic family must have the p-characteristics impinging on it. That is,

$$\lambda^p(q_l) > s > \lambda^p(q_r), \tag{41}$$

where s is the shock speed and λ^p is the pth eigenvalue evaluated at q_l and q_r — the states directly to the left and right of the shock respectively. Therefore, if the entropy solution to a Riemann problem contains a shock connecting the left state Q_{i-1} to the middle state, denoted Q_m, then

$$\lambda^1(Q_{i-1}) > \lambda^1(Q_m). \tag{42}$$

Similarly if a shock connects the right state Q_i to the middle state Q_m, then

$$\lambda^2(Q_m) > \lambda^2(Q_i). \tag{43}$$

If (42) or (43) is violated, then in fact a rarefaction connects the corresponding states in the entropy solution. As explained above, an entropy violating Riemann solution will not affect the numerical solution, except in the case of a transonic rarefaction. Therefore the only cases in which an entropy fix is required are when

$$\lambda^1(Q_{i-1}) < 0 < \lambda^1(Q_m) \tag{44}$$

or

$$\lambda^2(Q_m) < 0 < \lambda^2(Q_i), \tag{45}$$

which indicate a transonic rarefaction in the first or second families respectively.

It is easy to check $\lambda^1(Q_{i-1})$ and $\lambda^2(Q_i)$. However, with the f-wave approach, Q_m is never explicitly computed so we cannot simply evaluate $\lambda^1(Q_m)$ and $\lambda^2(Q_m)$ directly. In fact the f-wave approach is not equivalent to using a single value for Q_m, but two middle values, one to the left and one to the right of the cell interface. The approach we've taken is to instead compare the Roe speeds, $\hat{s}^1_{i-\frac{1}{2}}$ and $\hat{s}^2_{i-\frac{1}{2}}$, with the right and left speeds, $\lambda^1(Q_{i-1})$ and $\lambda^2(Q_i)$. The motivation for this approach is that the Roe speeds can serve as an estimate for the shock speeds, allowing us to estimate whether (41) is satisfied. For instance, to detect the presence of a transonic rarefaction in the second family, $u_i + \sqrt{gh_i}$ is compared to $\hat{s}^2_{i-\frac{1}{2}}$. If $u_i + \sqrt{gh_i} > \hat{s}^2_{i-\frac{1}{2}}$ then most likely the true Riemann solution has a rarefaction in the second family. Furthermore $\hat{s}^2_{i-\frac{1}{2}}$ serves as an estimate for the characteristic speed at the center of the rarefaction fan, $\hat{s}^2_{i-1/2} \approx \frac{1}{2}(\lambda^2(Q_m) + \lambda^2(Q_i))$ and hence $\lambda^2(Q_m) \approx 2s^2_{i-1/2} - \lambda^2(Q_i)$. Therefore if

$$2(\hat{s}^2_{i-\frac{1}{2}}) - (u_i + \sqrt{gh_i}) < 0 < u_i + \sqrt{gh_i}, \tag{46}$$

then the true Riemann solution is likely to have a transonic rarefaction in the second characteristic family. The second f-wave $\mathcal{Z}^2$ should then be split into two waves, one moving to the right the other to the left. A similar test is done for the first characteristic family.

The entropy fix when one state in the Riemann problem is initially dry is somewhat different, and also acts to ensure that the depth in the corresponding cell remains non-negative. As mentioned in the previous section, the true Riemann solution in such a case consists of a single rarefaction. The speeds of the edges of the rarefaction wave are given by quantities in the wet cell ($s^1 = u_{i-1} - \sqrt{gh_{i-1}}$ and $s^2 = u_{i-1} + 2\sqrt{gh_{i-1}}$ if $h_{i-1} > 0$ or $s^1 = u_i - 2\sqrt{gh_i}$ and $s^2 = u_i + \sqrt{gh_i}$ if $h_i > 0$). In the event of a single transonic rarefaction, $s^1 < 0 < s^2$, an entropy fix such that the f-waves simply carry an appropriate proportion of the true single wave is necessary. We simply apportion to each f-wave an amount based on the proportion of the rarefaction in each cell.

This approximate Riemann solver works quite robustly in the context of the first-order accurate Godunov's method, in the f-wave formulation with bathymetry source terms and dry states. Addition of the second-order correction terms (28) adds new potential problems when dry states are present, as these waves can lead to a negative depth even when the first order method would not. Standard wave limiters devised to avoid nonphysical oscillations may not completely avoid this problem, and so we have added an additional limiting procedure to maintain nonnegative depth.

The procedure is best illustrated by writing the second-order method as the sum of the first-order Godunov update

$$Q_i^G = Q_i^n - \frac{\Delta t}{\Delta x}(\mathcal{A}^+ \Delta Q_{i-1/2} + \mathcal{A}^- \Delta Q_{i+1/2}), \tag{47}$$

and the second-order correction fluxes (29),

$$Q_i^{n+1} = Q_i^G - \frac{\Delta t}{\Delta x}\left[\tilde{F}_{i+\frac{1}{2}} - \tilde{F}_{i-\frac{1}{2}}\right]. \tag{48}$$

Note that the correction flux at a cell interface takes mass away from one cell and adds mass to the adjacent cell, with the direction depending on the sign of its first component. The gross mass-flux out of the ith grid cell, due to these correction fluxes, is therefore

$$M_i = \left[\max(0, \tilde{F}^1_{i+\frac{1}{2}}) - \min(0, \tilde{F}^1_{i-\frac{1}{2}})\right]. \quad (49)$$

If $\Delta t M_i$ is larger than the mass present after the Godunov update, $\Delta x (Q_i^G)^1$, the correction fluxes could potentially create a negative depth in this cell. This is prevented by re-limiting the correction fluxes based on which cell they take mass away from

$$\tilde{F}_{i-\frac{1}{2}} \to \varphi_{i-\frac{1}{2}} \tilde{F}_{i-\frac{1}{2}}, \quad (50)$$

where

$$\varphi_{i-\frac{1}{2}} = \begin{cases} \min(1, \Delta x (Q_i^G)^1 / \Delta t M_i) & \text{if } \tilde{F}^1_{i-\frac{1}{2}} < 0 \\ \min(1, \Delta x (Q_{i-1}^G)^1 / \Delta t M_{i-1}) & \text{if } \tilde{F}^1_{i-\frac{1}{2}} > 0. \end{cases} \quad (51)$$

This procedure is consistent with the standard limiting strategy, in that second-order accuracy is achieved throughout most of the domain, and limiting only occurs where there are problematic features, such as shock waves or near dry-states.

6. Results for Benchmark Problem 1

The first benchmark problem consists of a linear sloping beach, $b(x,t) = -x$, and initially motionless water with a shoreline at $x = 0$. An incoming wave is induced by a non-zero initial surface elevation, $\eta(x,0)$, specified by data provided from $x = 0$ to 50000 meters in increments of 50 meters. We compare the resulting motion of the shoreline, the surface elevation, and the velocity field of our numerical solution with a provided analytical solution. The surface elevation and velocity data for the analytical solution were provided at three separate times, $t = 160$, 175, and 220 seconds. The position and velocity of the shoreline were provided for $t \in [0, 300]$ seconds.

We computed this problem using CLAWPACK, on a series of grids of varying resolution (from 500 to 5000 grid cells), each over a domain $x \in [-500, 50000]$ meters. The computational grid cells were clustered densely near the shoreline, using a piecewise linear mapping. This was required to efficiently capture the fine-scale motion of the shoreline, which occurred over a small fraction of the entire domain. Convergence to the analytical solutions was observed as the grids were refined.

Some of our computational results are shown with the analytical solutions in Figs. 1 and 2. We have chosen to show the results on a grid of

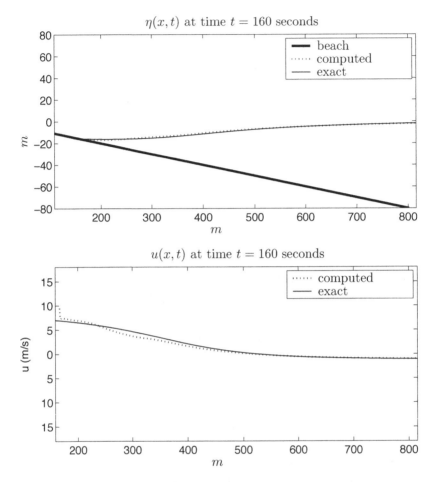

Fig. 1. Top: Water surface elevation at $t = 160s$, shown near the beach. Bottom: Velocity field at $t = 160s$ in the same region. Both computations were done on a 1000-point grid.

1000 points so that the numerical results are still distinguishable from the analytical solution. The surface elevation and the velocity field are shown at $t = 160s$ in Fig. 1. Note that the figures show only a small portion of the domain near the beach. Figure 2 shows the motion of the shoreline for the same computation. For additional results at other times and grid resolutions, see the website[15].

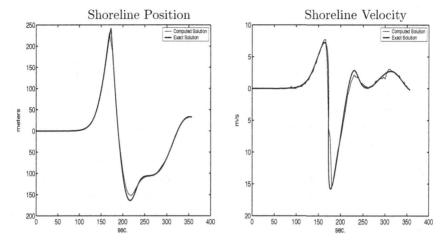

Fig. 2. Left: Position of the shoreline as a function of time, computed on a 1000-point grid. Right: Velocity of the shoreline for the same computation.

7. Extension to Two Space Dimensions

In two space dimensions the shallow water equations with bathymetry take the form

$$q_t + f(q)_x + g(q)_y = \psi(q, x, y), \tag{52}$$

where

$$q = \begin{bmatrix} h \\ hu \\ hv \end{bmatrix}, \quad f(q) = \begin{bmatrix} hu \\ hu^2 + \tfrac{1}{2}gh^2 \\ huv \end{bmatrix}, \quad g(q) = \begin{bmatrix} hv \\ huv \\ hv^2 + \tfrac{1}{2}gh^2 \end{bmatrix}, \tag{53}$$

and the source term is $\psi = \psi_1 + \psi_2$ with

$$\psi_1(q, x, y) = \begin{bmatrix} 0 \\ -ghB_x(x,y) \\ 0 \end{bmatrix}, \quad \psi_2(q, x, y) = \begin{bmatrix} 0 \\ 0 \\ -ghB_y(x,y) \end{bmatrix}, \tag{54}$$

The general form of the wave-propagation algorithm is now

$$Q_{ij}^{n+1} = Q_{ij} - \frac{\Delta t}{\Delta x}(\mathcal{A}^+ \Delta Q_{i-1/2,j} + \mathcal{A}^- \Delta Q_{i+1/2,j})$$

$$- \frac{\Delta t}{\Delta y}(\mathcal{B}^+ \Delta Q_{i,j-1/2} + \mathcal{B}^- \Delta Q_{i,j+1/2}) \tag{55}$$

$$- \frac{\Delta t}{\Delta x}(\tilde{F}_{i+1/2,j} - \tilde{F}_{i-1/2,j}) - \frac{\Delta t}{\Delta y}(\tilde{G}_{i,j+1/2} - \tilde{G}_{i,j-1/2}).$$

The fluctuations $\mathcal{A}^{\pm}\Delta Q$ and $\mathcal{B}^{\pm}\Delta Q$ are determined by solving one-dimensional Riemann problems normal to each cell interface, using the f-wave formulation described above to incorporate the appropriate portion of the source term. In the x-direction, for example, we solve the Riemann problem $q_t + f(q)_x = \psi_1$. The solution to this Riemann problem is exactly the same as the one-dimensional case with the addition of a contact discontinuity wave that passively advects the jump in the orthogonal velocity v. Using only these terms in (55) without the correction fluxes $\tilde{F}$ or $\tilde{G}$ gives a two-dimensional generalization of Godunov's method sometimes called donor-cell upwind. A better first-order method (corner transport upwind) is obtained by splitting the waves in each one-dimensional Riemann solution into waves moving in the transverse direction, so that wave motion oblique to the grid is more properly modeled. In CLAWPACK this requires the specification of a "transverse Riemann solver" that takes as input a fluctuation, say $\mathcal{A}^+\Delta Q$, and returns a splitting of this vector into up-going and down-going portions that affect the fluxes $\tilde{G}_{i,j+1/2}$ and $\tilde{G}_{i,j-1/2}$ respectively. This is typically based on the eigenvalues and eigenvectors of some approximate Jacobian matrix $g'(q)$ in the transverse direction, and is described in more detail for the shallow water equations in LeVeque[14]. Second-order correction terms can also be incorporated as in one space dimension, based on the waves obtained from the one-dimensional Riemann solution normal to each cell edge.

The presence of dry states leads to new complications when the algorithm is extended to two space dimensions. A wave moving transversely into a dry or nearly dry cell can produce a negative depth unless special care is taken by incorporating a more fully multidimensional limiter. Currently we do not use transverse propagation or second-order correction terms in two dimensions and are further developing this approach. Our goal is to ultimately be able to use transverse propagation and high-resolution correction terms for dry state problems in two dimensions. However, even without these terms reasonable results are obtained, as demonstrated in the next section.

8. Results for Benchmark Problem 2

In the second benchmark problem, we compare our computational solutions to laboratory data collected from a wave tank experiment. The wave tank was built as a 1:400 scale model, approximating coastline bathymetry near Monai, Japan, a region that suffered inundation from the 1993 Okushiri

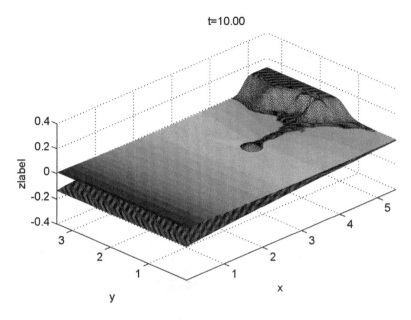

Fig. 3. Three-dimensional view of the numerical solution at one time.

tsunami. In the experiment, an incoming wave was induced by mechanical wave paddles along one side of the tank, and the resulting motion of the surface elevation was measured by gages at three separate locations. Additionally, a movie was made showing the tank from overhead during the experiment. This data and movie are available at the workshop website.

We computed this problem with CLAWPACK, originally using a uniform single grid of 275×175 cells. The computational domain modeled a subregion of the tank, $(x, y) \in [0, 5\text{m}] \times [0, 3.5\text{m}]$, which includes an inlet region that experienced large wave runup in the tsunami. We used the provided data that specifies the measured surface elevation of an incoming wave along one side of a subregion of the tank, $x = 0$, for the period $t = [0, 22.5\text{s}]$. (Note: the computational domain is rotated in the figures to match the orientation of the movie, so x runs upward and y from left to right.)

The computed surface elevation is shown at various times in Figs. 3 through 6. Figure 3 shows a three-dimensional view of the computed solution at one time, Fig. 4 shows the solution during the primary runup period, and Fig. 5 compares the solution with several close-up snapshots from the wave tank movie, demonstrating similarities in the zone where the maximum runup occurred.

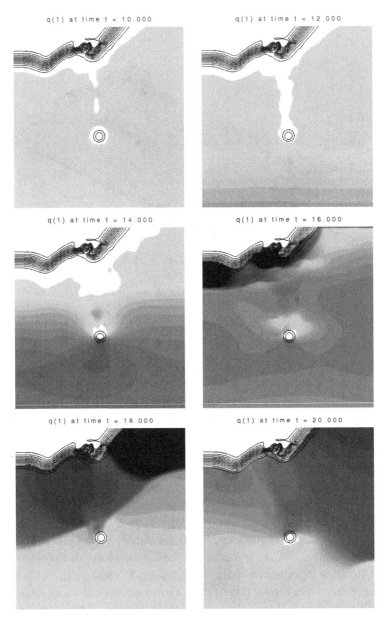

Fig. 4. The computed solution during the primary runup of the wave. The primary runup occurred in the first 30 seconds. Contour lines show the topography that was initially above the water surface, including an island. Gray scale indicates elevation above the original water surface.

High-Resolution Finite Volume Methods for the Shallow Water Equations 65

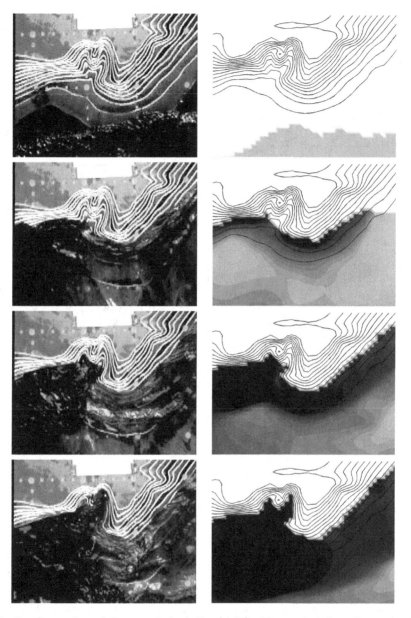

Fig. 5. Comparison of the numerical solution (right) with snapshots from the overhead movie of the laboratory experiment (left). Frames 11, 26, 41, 56 from the movie are shown and computed results are shown at corresponding 30 second intervals. (The movie shows 30 frames per second.)

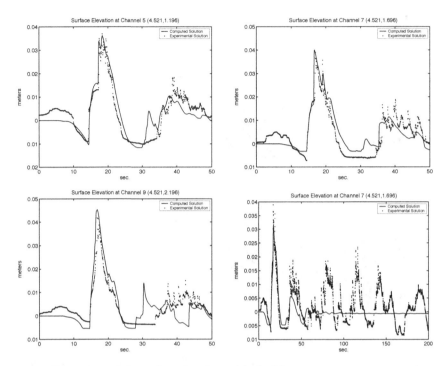

Fig. 6. Comparison of the surface elevation, with the laboratory measurements at the three wave gages (Channels 5, 7, 9). The numerical solution is comparable during the primary runup in the first 50 seconds. The bottom right figure shows the comparison for the duration of the laboratory measurements, at one of the gages (Channel 5). Sustained oscillations in the laboratory measurements are not seen in the numerical solution, as discussed in the text. The other gages showed similar patterns at later times.

Laboratory measurements at three gage locations were provided for the workshop, denoted Channel 5, 7, and 9. This data and the numerical solution (surface elevation as a function of time) at these three locations are shown in Fig. 6. The numerical solution was comparable to the laboratory measurements for the first 50 seconds of the experiment, during which the primary runup of the wave and several reflected waves are seen. For the last 150 seconds of the experiment, the wave tank measurements exhibited large oscillations that were not evident in the numerical solution. This discrepancy is shown in the last graph of Fig. 6. Based on discussions at the workshop, it is believed that this is due to the fact that the wave maker in the experiment did not generate a perfectly clean wave. The incoming wave data provided for the benchmark problem was only over 22.5 seconds, and was followed by other waves in the wave tank for which no data was provided.

9. Adaptive Mesh Refinement and Large-Scale Tsunami Propagation

In two-dimensional calculations over large spatial domains it may be crucial to use nonuniform grids to achieve the desired resolution in some regions without an excessive number of grid cells overall. For tsunami modeling there are two types of nonuniformity that may be desirable. First, we may want to have a finer grid near the shore where the details of the runup must be computed, or in other regions where the bathymetry has a large impact on the wave propagation behavior and eventual runup. In principle this can be achieved with "static refinement" of some spatial regions relative to others, though ideally computations on the finer grid would only be done at times when the wave is present. Second, we may want to use an adaptively refined grid in which the region of refinement moves along with the tsunami to provide good resolution of the wave at all times, with a minimal number of grid points in regions where nothing is happening.

Since the Catalina workshop took place, the tragic Sumatra tsunami of December 26, 2004 has prompted increased study of the all aspects of tsunamis. Our own efforts in recent months have focused on developing an adaptive mesh refinement code that works well on the scale of the Indian Ocean (or other oceans), with the ability to capture tsunami propagation across the ocean and couple it to the study of runup on much smaller scales along particular stretches of coastline. In the course of this work we have also improved the Riemann solver beyond what is described in this paper to make it even more robust. The new version no longer requires increasing the wave speeds above the Einfeldt speeds, and instead is based on introducing additional waves in the Riemann solution using ideas from LeVeque and Pelanti[16]. This work will be reported in more detail elsewhere. See the webpage[15] for pointers.

The CLAWPACK software includes AMRCLAW, an adaptive mesh refinement version of the code developed with Marsha Berger and described in more detail in Berger and LeVeque[2]. More recently the CLAWPACK formulation has also been incorporated into ChomboClaw by Donna Calhoun at the University of Washington, allowing the Chombo package[4] of C++ routines for adaptive refinement to be applied. This package uses similar algorithms to AMRCLAW but has additional features such as an MPI implementation with load balancing for adaptive refinement on parallel computers, and support for implicit algorithms that may be needed if dissipative or dispersive terms are added to the equations.

A rectangular grid is refined by covering a portion of the domain by one or more rectangular patches of grids that are finer by a factor of k, some integer. This process can be repeated recursively, with each grid level further refined by grid patches at a higher level or refinement. The maximum number of levels is specified along with the refinement factor at each level. The grid patches are chosen by flagging grid cells at each level that "need refining" (see below) and then clustering these cells into a set of rectangular patches that cover these cells and a limited number of other cells. This is done by solving an optimization problem (using the algorithm of Berger and Rigoutsos[3]) that takes into account the trade-off between refining too many cells unnecessarily and creating too many grid patches, since there is some overhead associated with applying the algorithm on each patch. The boundary data (ghost cell values) on a patch must be generated by space-time interpolation from data on the coarser grid. A time step of length Δt is first taken on the coarse grid, boundary data is then generated from the fine grid patches, and k' time steps of length $\Delta t/k'$ are then taken on the fine grids to reach the same time. For most AMR applications on hyperbolic problems we take $k' = k$, refining in time by the same factor as in space in order to maintain a comparable Courant number at all levels. However, for tsunami propagation and runup applications where we only have the finest grids near the shore, it may be desirable to refine in time by a smaller factor $k' < k$. Recall that the wave speeds are roughly $\sqrt{gh}$, which differs by more than an order of magnitude between the deep ocean and the coastal regions. The smaller wave speed near shore allows a larger time step.

After updating to the same time on the fine grids, the coarse grid solution on any grid cell covered by the finer grid is then replaced by the average of the fine grid values in this cell in order to transfer the more accurate solution to the coarse grid. Additional modifications to the values are needed near the patch edge to maintain global conservation as described in Berger and LeVeque[2]. This is done recursively at each level. While the inner workings are somewhat complicated, the general formulation in AMRCLAW allows extension of most CLAWPACK computations directly to adaptively refined grids. The computational time required for the overhead associated with multiple grids is often negligible compared to the savings achieved by concentrating fine grids only where needed.

We can flag the cells that "need refinement" however we wish. For the calculation presented below, we have allowed some refinement everywhere based on a measure of the variation in the solution, so that the propagating wave is well resolved. In addition, we allow additional levels of refinement near the shore in particular regions of interest, though only once the wave is approaching. Regridding is performed every few time steps to modify the region of refinement as the wave propagates.

In the course of this work, we ran into several difficulties at the interfaces between grids that have not been observed in other applications of the AMR codes. These arise from the representation of the bathymetry and shore on a Cartesian grid. One problem, for example, is that a coarse grid cell that is dry (if the bathymetry value in this cell is above sea level) may be refined into some cells that are above sea level and others below. Even though $h = 0$ on the coarse cell we cannot initialize h to zero on all the fine cells without generating nonphysical wave motion. The fine cells below sea level must be initialized with a positive h to maintain the constant sea level. This problem and related difficulties could only be solved by some substantial reworking of the AMRCLAW code. The result is a special-purpose program that incorporates these algorithmic modifications and can now be applied to many tsunami scenarios. It is currently being tested by comparing predictions with measurements made at various places around the Indian Ocean in the wake of the 26 December 2004 Sumatra earthquake. Some preliminary results are shown in Fig. 7, and additional results and movies are linked from the webpage[15] and will be reported more fully in future publications.

The top two frames of Fig. 7 show the Bay of Bengal at two early times. A coarse grid is used where nothing is yet happening and the grid cells are shown on this "Level 1 grid", which has a mesh width of one degree (approximately 111 km). The rectangular region where no grid lines are shown is a Level 2 grid with mesh width 8 times smaller, about 14 km. Red represents water elevation above sea level, dark blue is below the undisturbed surface. Figure 7(c) shows a zoomed view of the southern tip of India and Sri Lanka at a later time. The original Level 1 grid is still visible along the left-most edge, but the rest of the region shown has been refined by a Level 2 grid. Due north from Sri Lanka, along the coast of India, there is a region near Chennai where two additional levels of refinement have been allowed at this stage in the computation. The grid lines on Level 3 are not shown; the mesh width on this level is about 1.7 km, a factor of 8 finer than Level 2. A Level 4 grid is also visible in the center of this region and appears as a small black rectangle.

(a) Time 01:07:10 (550 seconds)   (b) Time 01:34:40 (2200 seconds)

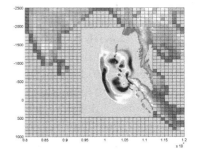

(c) Time 03:13:12.5 (8112.5 seconds)   (d) Time 03:15:30 (8250 seconds)

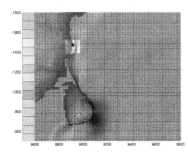
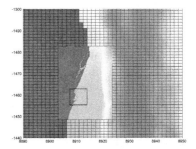

(e) Time 03:24:40 (8800 seconds)   (f) Time 03:35:30 (9450 seconds)

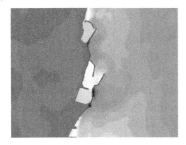
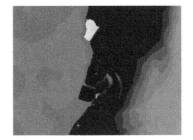

Fig. 7. Propagation of the 26 December 2004 tsunami across the Indian Ocean, using adaptive mesh refinement with refinement by a factor of 4096 from the coarsest grid shown in (a)–(b) to the finest grid shown in (d)–(f). The latter figures show zoomed views of the region near the harbors of Chennai, India. Times are GMT (and seconds since initial rupture at 0:58 GMT). See the text for more details.

Figure 7(d) shows a further zoomed view of the coast near Chennai. In this figure the grid lines show the Level 3 grid. The Level 4 grid is refined by a factor of 64 relative to the Level 3 grid, so the mesh width is about 27 meters. Grid lines on this finest level are not shown. The rectangle in this figure shows a region that is expanded in Figs. 7(e) and 7(f) to show the two harbors of Chennai. The fine-scale bathymetry used in this computation was obtained by digitizing navigational charts. In this simulation the commercial harbor to the south is inundated, with the tsunami overtopping the surrounding sea wall, while the fishing harbor to the north is largely untouched. This appears to agree with what was actually observed, although we are still investigating this. Moreover, we do not yet have sufficiently accurate data on the height of the sea walls enclosing each harbor. One data point known accurately from tide gage data is the arrival time for the initial wave in Chennai, which was at 9:05 local time (3:35 GMT). This is well matched by our simulation: this is essentially the time shown in Fig. 7(f).

A more careful study of this region will be performed in the future and presented elsewhere, including comparisons with runup data collected by Harry Yeh as part of the Tsunami Survey Team that visited this area in February, 2005. See the webpage[15] for more recent results and movies of the simulations. In the future we plan to also compare with data collected by other teams at various other points around the Indian Ocean. We will also make our computer code available to the community for others to use, in a form that should allow the local bathymetry from other regions to be easily incorporated. See the webpage for more details.

Adaptive mesh refinement is crucial for this simulation. The calculation shown here was run on a single-processor 3.2 GHz PC under linux. Figure 7(c) was obtained after about 15 minutes of running time, Fig. 7(d) about 2.5 hours later, indicating that most of the grid cells are concentrated on the Level 4 grid, which is introduced only when the wave approaches the shore in this one region. Were it possible to use the finest grid over the entire domain, the result of Fig. 7(c) would require more than 4000 years of computing on the same machine.

10. Conclusions and Future Work

Finite volume methods using an approximate Riemann solver devised to deal with bathymetry and dry states have been successfully used for Benchmark Problems 1 and 2. Benchmark problems 3 and 4 at the workshop model landslide-induced tsunamis, and involve bathymetry that changes with time. In principle our code can handle this situation but we have not yet done extensive tests of this.

Adaptive mesh refinement is crucial for large scale problems and we are further developing and testing the AMR version of our method. We are also working on formulating this method on the full earth, by switching to latitude-longitude coordinates on the sphere and calculating on a computational rectangle in these coordinates using periodic boundary conditions in longitude. The AMR code is already capable of handling this situation and expect to soon be able to study the global propagation and compare with tide gage data (see, for example, Titov et al.[18]) that is available from many points around the world following the 26 December 2004 event.

Acknowledgments

This work was supported in part by NSF grants DMS-0106511 and CMS-0245206, and by DOE grant DE-FC02-01ER25474. The authors would like to thank Marsha Berger and Donna Calhoun for assistance with the adaptive refinement aspect of this work.

References

1. D. Bale, R. J. LeVeque, S. Mitran, and J. A. Rossmanith, A wave-propagation method for conservation laws and balance laws with spatially varying flux functions, *SIAM J. Sci. Comput.*, **24** (2002), pp. 955–978.
2. M. J. Berger and R. J. LeVeque, Adaptive mesh refinement using wave-propagation algorithms for hyperbolic systems, *SIAM J. Numer. Anal.*, **35** (1998), pp. 2298–2316.
3. M. J. Berger and I. Rigoutsos, An algorithm for point clustering and grid generation, *IEEE Trans. Sys. Man & Cyber.*, **21** (1991), pp. 1278–1286.
4. P. Colella *et al.*, CHOMBO software, http://seesar.lbl.gov/anag/chombo/, 2005.
5. B. Einfeldt, On Godunov-type methods for gas dynamics, *SIAM J. Num. Anal.*, **25** (1988), pp. 294–318.
6. B. Einfeldt, C. D. Munz, P. L. Roe, and B. Sjogreen, On Godunov type methods near low densities, *J. Comput. Phys.*, **92** (1991), pp. 273–295.
7. L. Gosse, A well-balanced flux-vector splitting scheme designed for hyperbolic systems of conservation laws with source terms, *Comput. Math. Appl.*, **39** (2000), pp. 135–159.
8. J. M. Greenberg, A. Y. LeRoux, R. Baraille, and A. Noussair, Analysis and approximation of conservation laws with source terms, *SIAM J. Numer. Anal.*, **34** (1997), pp. 1980–2007.

9. A. Harten, P. D. Lax, and B. van Leer, On upstream differencing and Godunov-type schemes for hyperbolic conservation laws, *SIAM Review*, **25** (1983), pp. 35–61.
10. P. Jenny and B. Müller, Rankine-Hugoniot-Riemann solver considering source terms and multidimensional effects, *J. Comp. Phys.*, **145** (1998), pp. 575–610.
11. A. Kurganov and D. Levy, Central-upwind schemes for the Saint-Venant system, *Math. Model. and Numer. Anal.*, **36** (2002), pp. 397–425.
12. R. J. LeVeque, Wave propagation algorithms for multi-dimensional hyperbolic systems, *J. Comput. Phys.*, **131** (1997), pp. 327–353.
13. R. J. LeVeque, Balancing source terms and flux gradients in high-resolution Godunov methods: The quasi-steady wave-propagation algorithm, *J. Comput. Phys.*, **146** (1998), pp. 346–365.
14. R. J. LeVeque, *Finite Volume Methods for Hyperbolic Problems* (Cambridge University Press, 2002).
15. R. J. LeVeque and D. L. George, http://www.amath.washington.edu/~rjl/catalina04/.
16. R. J. LeVeque and M. Pelanti, A class of approximate Riemann solvers and their relation to relaxation schemes, *J. Comput. Phys.*, **172** (2001), pp. 572–591.
17. P. L. Roe, Approximate Riemann solvers, parameter vectors, and difference schemes, *J. Comput. Phys.*, **43** (1981), pp. 357–372.
18. V. Titov, A. B. Rabinovich, H. O. Mofjeld, R. E. Thomson, and F. I. Gonzalez, The global reach of the 26 December 2004 Sumatra tsunami, *Science*, **309** (2005), pp. 2045–2048.
19. E. F. Toro, *Shock-Capturing Methods for Free-Surface Shallow Flows* (Wiley and Sons Ltd., UK, 2001).

CHAPTER 3

SPH MODELING OF TSUNAMI WAVES

Benedict D. Rogers and Robert A. Dalrymple[*]

Department of Civil Engineering, The Johns Hopkins University
3400 N Charles St, Baltimore, Maryland, 21218, U.S.A.
E-mail: [*]rad@jhu.edu

Smoothed particle hydrodynamics (SPH) is used to simulate both two and three-dimensional cases of tsunami generation and propagation. A combined SPH-LES type scheme or sub-particle scale (SPS) scheme is used to describe viscous effects. Results for 2-D compare well with other nonlinear numerical models in the case of a 2-D landslide-generated tsunami. The application of the 2-D model to the 3-D landslide problem is shown to be inappropriate, necessitating a fully 3-D model. However, the resolution required for 3-D is beyond the current capabilities of the JHU-SPH code necessitating high performance parallel computing.

1. Introduction

While the mechanisms that generate the tsunami flows are generally understood, the ability to predict the flows they produce at coastlines still represents a formidable challenge due to the complexities of coastline formations and the presence of numerous coastal structures that interact and alter the flow. Indeed, overland flows kill more people alone than any other aspect of tsunamis but are still not well understood or accurately predictable. Numerous experimental investigations have focused on different instances of tsunami runup such as circular islands (Briggs *et al.* 1994). However, the accurate prediction of runup onto real coastlines with structures and human habitation still remains incomplete.

Theoretical approaches to the problem are extremely difficult to apply due to strong nonlinearities of tsunamis, the essential three-dimensionality of the flow, and the dramatic turbulent flow that results from a breaking tsunami. A few numerical models, such as shock-capturing schemes (Toro *et al.* 1999) and Reynolds-Averaged Navier-Stokes (RANS) (Lin *et al.* 1999), have been used to predict the flow of tsunamis and runup/rundown in highly simplifed cases. However, as yet, such numerical schemes have been unable to capture the flow following the passage of a tsunami when it breaks onto a beach and interacts with structures in a destructive manner. Furthermore, the use of depth-averaged techniques, which also includes Boussinesq models (Lynett *et al.* 2002), does not adequately predict the full structure of the flow, which is evidently important given that upon landfall tsunamis are propagating over dry land and are often interacting with structures on the same scale as the depth. Hence, fully 3-D numerical models are essential.

To date however, the numerical study of three-dimensional phenomena has been very difficult due to their complexity, the physical scales of the flow regimes, and the presence of the free surface. As part of this development, 3-D boundary element numerical models are in development to predict the initial propagation of the surface disturbance (Grilli *et al.* 2002). Techniques such as Eulerian grid-based Large Eddy Simulation (LES) and RANS (Christensen and Deigaard 2001, Lin *et al.* 1999) offer one approach to solve for these flows since they can resolve important components of the flow such as mean flow and turbulence. Alternatively, the development of meshfree methods for CFD (Li and Liu 2002) offers a different realizable approach to tackle the simulation of such difficult problems. Of the meshfree techniques developed over the past decade, Smoothed Particle Hydrodynamics (SPH) has been the most popular and successful when applied to free-surface hydrodynamics.

Smoothed Particle Hydrodynamics (SPH) is a Lagrangian technique originally developed for the simulation of non-axisymmetric problems in astrophysics that predicts the motion of discrete particles with time. Since its initial development in the late 1970s by Gingold and Monaghan (1977) and Lucy (1977) the technique has been extended to many other applications including free-surface hydrodynamics, primitive impact-fracture problems (Randles and Libersky 1996) and modelling solid-solid impact (Johnson and Beissel 1996).

The core of the SPH formalism lies in the discrete summation over disordered points as an approximation to integrals. A major attraction of the SPH technique is that the need for fixed computational grids is removed when calculating spatial derivatives. Instead, estimates of derivatives are provided by analytical expressions. Thus, it has a distinct advantage over other grid-based methods of solving for turbulent tsunami propagation problems in that there is no meshing required. Furthermore, the domain can be multiply connected, and there is no need for special treatment of the free surface; thus wave breaking, and other flows that result in fluid separation, are modeled as readily as other flows.

SPH has been used to describe a variety of free-surface flows including solitary wave propagation over a planar beach (Monaghan and Kos 1999), plunging breakers (Tulin and Landrini 2000), solid bodies impacting on the surface (Monaghan 2000) and dam break simulations (Monaghan 1994). The method predicts fluid pressure, velocities, energy and particle trajectories for many types of flows making it ideal for identifying/elucidating formation mechanisms of complicated flow phenomena that were hitherto intractable. Additionally, using a similar particle technique known as the Moving Particle Semi-implicit (MPS) method, Koshizuka *et al.* (1998) and Gotoh *et al.* (2002) present two dimensional results for the movement of passively floating bodies in the surf zone and an open channel, respectively. Such an approach can be incorporated into SPH allowing one to examine the effect of tsunami propagation into coastal areas that have debris within its domain and whose buildings are subject to destruction.

At Johns Hopkins University (JHU), Dalrymple and Knio (2001) and Dalrymple *et al.* (2001) have shown that SPH has great potential for a wide range of coastal wave propagation problems such as the greenwater overtopping of offshore platforms, the waves generated by subaerial landslide impact, and the impact of waves on structures. Gómez-Gesteira and Dalrymple (2004) have been the first to develop a three-dimensional free-surface SPH flow model applying it to the impact and hydrodynamics of a large wave striking and flowing around a simple square building — essentially a simple model of a tsunami at the shoreline. The model is able to track particles, determine forces on structures, and replicate velocities accurately when compared to laboratory data. Thus, particle methods now also represent a viable method for simulating flow around buildings and complicated fluid-structure interaction problems. More recently, Panizzo and Dalrymple (2004) applied SPH to the two-dimensional modeling of tsunamis using the waves generated by underwater sliding block experiments as verification.

In this paper, Sec. 2 introduces the basic SPH discretization. This is followed by a description of an LES-type sub-particle-scale formulation that enhances the basic SPH formulation to model turbulent flow features. The formulation is then completed by a presentation of the boundary conditions used within the JHU-SPH scheme. Section 3 presents the numerical results, which are compared with an analytical solution for tsunami generation due to a two-dimensional landslide, and experimental data for tsunami generation due to a three-dimensional landslide. Finally, Sec. 4 draws some conclusions and looks at the future of applying SPH to the simulation of tsunamis.

2. Numerical Discretization

2.1. *Basic SPH Formulation*

Smoothed Particle Hydrodynamics (SPH) is based on the approximate representation of continuous interpolations or integrals by a discrete particle representation. In the SPH formalism, the value of a flow quantity, f, at a position vector $\mathbf{x}$ is approximated by

$$f(\mathbf{x}) = \sum_j f_j W(\mathbf{x} - \mathbf{x}_j) V_j , \qquad (2.1)$$

where V_j is the volume of the jth particle located at $\mathbf{x}_j$ with scalar quantity f_j, and $W(\mathbf{x} - \mathbf{x}_j)$ is the weighting function referred to as the smoothing kernel. Prior to simulation, the kernel is specified by an analytical expression making its evaluation in Eq. (2.1) straightforward. In all results presented, a quadratic kernel has been used for computational speed and to avoid the problems associated with inflexion points in the derivatives of spline kernels (Liu *et al.* 2003):

$$W(\mathbf{x}-\mathbf{x}_j) = \begin{cases} C_N\left(\tfrac{1}{4}q^2 - q + 1\right) & 0 \le q \le 2 \\ 0 & \text{otherwise}, \end{cases} \quad (2.2)$$

where $q = r/h$, and $r = |\mathbf{x}-\mathbf{x}_j|$. In 1-D, $C_N = C_1 = 3/4h$, in 2-D $C_N = C_2 = 3/(2\pi h^2)$ and in 3-D $C_N = C_3 = 15/(16\pi h^3)$. The kernel has a characteristic smoothing length, h, which defines the region of influence, and is preset to be $1.3\Delta x$ for the duration of all simulations where Δx is the initial particle spacing.

One of the attractions of SPH is that function derivatives can be expressed as another summation by simply using a derivative of the smoothing kernel, i.e.

$$\frac{\partial f(\mathbf{x}_i)}{\partial x_i} = \sum_j (f_j - f_i) \frac{\partial W(\mathbf{x}_i - \mathbf{x}_j)}{\partial x_i} V_j. \quad (2.3)$$

This avoids complicated expressions for calculating derivatives since an analytical expression is known for the specified kernel. Expression (2.3) also ensures that there is zero divergence for a uniformly constant field.

When simulating fluid flow, SPH is based on a discrete particle representation of the Navier-Stokes equations in a Lagrangian formulation

$$\frac{d\rho}{dt} = -\rho \nabla \cdot \mathbf{u}, \quad (2.4a)$$

$$\frac{d\mathbf{u}}{dt} = -\frac{\nabla P}{\rho} + \mathbf{g} + \frac{1}{\rho}(\nabla \cdot \mu \nabla)\mathbf{u}, \quad (2.4b)$$

where ρ is density, μ is laminar viscosity, $\mathbf{u}$ is velocity, P is pressure, $\mathbf{g}$ is gravity and t is time. Following Monaghan (1994), to avoid solving the Poisson equation every timestep for an incompressible system in simulations involving water, the pressure is given by an approximate equation of state (Batchelor 1967)

$$P = B\left[\left(\frac{\rho}{\rho_o}\right)^\gamma - 1\right], \quad (2.5)$$

where $\rho_o = 1000 \text{kg/m}^3$ is a reference density for water, $\gamma = 7$ is a polytropic index, and the factor B is chosen so that water is virtually incompressible where the speed of sound, $c = \sqrt{\partial P/\partial \rho}$, does not dictate a time step that is prohibitively small, that is $c \approx 10 u_{max}$. Written in particle form, Eqs. (2.4) become

$$\frac{d\rho_i}{dt} = \sum_j m_j (\mathbf{u}_i - \mathbf{u}_j) \cdot \nabla_i W_{ij}, \qquad (2.6a)$$

$$\frac{d\mathbf{u}_i}{dt} = -\sum_j m_j \left(\frac{P_i}{\rho_i^2} + \frac{P_j}{\rho_j^2} + \Pi_{ij} \right) \nabla_i W_{ij} + \mathbf{g}, \qquad (2.6b)$$

where m_j is the mass of the jth particle, $\nabla_i W_{ij} = \nabla_i W(\mathbf{x} - \mathbf{x}_j)$, the first subscript i referring to the derivative of W with respect to the coordinates of particle i, and Π_{ij} is an artificial viscosity term. The artificial viscosity term (see Monaghan 1992) is used to account for the viscous terms in the governing equations. For free-surface hydrodynamics, the artificial viscosity term also prevents the simulation from becoming unstable and is given by (Monaghan 1992)

$$\Pi_{ij} = -\frac{\alpha \mu_{ij} \bar{c}_{ij}}{\bar{\rho}_{ij}}, \qquad (2.7)$$

where α is an empirical coefficient (usually taken as $0.01 - 0.1$), $\bar{c}_{ij} = \frac{1}{2}(c_i + c_j)$, $\bar{\rho}_{ij} = \frac{1}{2}(\rho_i + \rho_j)$ and

$$\mu_{ij} = \frac{h \mathbf{u}_{ij} \cdot \mathbf{x}_{ij}}{r_{ij}^2 + \eta^2}. \qquad (2.8)$$

Here, $\mathbf{u}_{ij} = \mathbf{u}_i - \mathbf{u}_j$, $r_{ij} = |\mathbf{x}_i - \mathbf{x}_j|$ and $\eta^2 = 0.01 h^2$ to prevent singularities.

2.2. Sub-Particle Scale (SPS) Turbulence Model

It is now recognized that the original formulation of viscous effects in SPH using the conventional form in Eq. (2.7) suggested by Monaghan (1992) does not provide an adequate or complete description of the physics in free-surface hydrodynamics. The empirical coefficient,

α, is needed for numerical stability for free-surface flows, but, in practice, is too dissipative. Thus, in recent years there has been a steady shift towards improved formulations. An important step in this direction was the work by Morris *et al.* (1997) who developed an improved formulation to account for laminar stresses

$$\frac{1}{\rho}(\nabla \cdot \mu \nabla)\mathbf{u}\bigg|_i = \sum_j m_j \frac{v_o(\rho_i + \rho_j)}{\overline{\rho}_{ij}^2} \frac{\mathbf{r}_{ij} \cdot \nabla_i W_{ij}}{r_{ij}^2 + 0.01h^2} \mathbf{u}_{ij}, \qquad (2.9)$$

where $v_o = 10^{-6}$ m^2/s is the kinematic viscosity. This avoids using second-order derivatives of the smoothing kernel as these can be inaccurate under certain circumstances (Morris *et al.* 1997).

In other areas of CFD, the concept of Large-Eddy Simulation (LES) has proved useful in capturing coherent turbulent structures within turbulent flows (see Meneveau and Katz 2000 for a review of LES models). In LES, the Navier-Stokes equations are filtered spatially over a length scale similar to the grid size to account for turbulent motions occurring on scales smaller than the resolved grid resolution. Closure submodels are then required to model the sub-grid-scale (SGS) turbulence that completes the system of equations.

Hence, given the unphysical damping behavior associated with the artificial viscosity approach, the next logical step has been to incorporate the ideas behind LES into particle simulations. Gotoh *et al.* (2001) were the first to introduce the enhancement of an LES-type scheme within their incompressible MPS technique. Shortly after, Lo and Shao (2002) became the first to implement an LES-type filtering within their incompressible version of SPH (I-SPH). This was followed by Shao and Gotoh (2004) and Gotoh *et al.* (2004) who successfully applied their I-SPH-LES scheme to wave-structure interaction problems with a partially submerged cut-off wall.

In this work, we solve the compressible flow equations, and accordingly, we are required to include the effects of compressibility in the development of an SPH-LES scheme. This is achieved by Favre-averaging ($\tilde{f} = \overline{\rho f}/\overline{\rho}$) (where '–' denotes a particle-scale quantity). This is a useful form of spatial averaging that does not introduce extra terms into the continuity equation. Applying an implicit (or flat-top) spatial-filter over the particle-scale governing equations, Eqs. (2.4) become (Yoshizawa 1986)

$$\frac{d\bar{\rho}}{dt} = -\bar{\rho}\nabla\cdot\tilde{\mathbf{u}}, \qquad (2.10a)$$

$$\frac{d\tilde{\mathbf{u}}}{dt} = -\frac{1}{\bar{\rho}}\nabla\bar{P} + \mathbf{g} + \frac{1}{\bar{\rho}}(\nabla\cdot\bar{\rho}v_o\nabla)\tilde{\mathbf{u}} + \frac{1}{\bar{\rho}}\nabla\cdot\vec{\tau}, \qquad (2.10b)$$

where $\vec{\tau}$ is the sub-particle scale (SPS) stress tensor with elements (in tensor notation):

$$\tau_{ij} = \bar{\rho}\left(2v_t\tilde{S}_{ij} - \tfrac{2}{3}\tilde{S}_{kk}\delta_{ij}\right) - \tfrac{2}{3}\bar{\rho}C_I\bar{\Delta}^2\delta_{ij}. \qquad (2.11)$$

Following Blin et al. (2002), $C_I = 0.00066$. The strain tensor, $\tilde{S}_{ij}$, is given by

$$\tilde{S}_{ij} = \frac{1}{2}\left(\frac{\partial\tilde{u}_i}{\partial x_j} + \frac{\partial\tilde{u}_j}{\partial x_i}\right). \qquad (2.12)$$

As the name suggests, sub-particle scale (SPS) schemes are very similar to sub-grid scale (SGS) schemes used in Eulerian grid-based LES schemes. The final stress-gradient term in Eq. (2.10) represent the effect of turbulent motions occuring on length scales smaller than the particle-scales of the resolved flow. Herein, it is accepted that more sophisticated viscosity models have been developed for grid-based LES schemes, but, in this work the standard Smagorinsky model (1963) for the eddy viscosity appears sufficient. This is given by

$$v_t = (C_s\Delta l)^2 |\bar{S}|, \qquad (2.13)$$

where the Smagorinsky constant, $C_s = 0.12$, Δl is the initial particle spacing and the local strain rate is given by

$$|\bar{S}| = \left(2\bar{S}_{ij}\bar{S}_{ij}\right)^{1/2}. \qquad (2.14)$$

Within our SPH scheme we discretize the SPS stresses according to the symmetric formulation given by Lo and Shao (2002)

$$\left.\frac{1}{\rho}\nabla\cdot\vec{\tau}\right|_i = \sum_j m_j\left(\frac{\vec{\tau}_i}{\rho_i^2} + \frac{\vec{\tau}_j}{\rho_j^2}\right)\cdot\nabla_i W_{ij}. \qquad (2.15)$$

2.3. Boundary Conditions

Boundaries are described by discrete particles that interact with the interior water particles. In the test cases presented in Sec. 3, open boundaries are modeled by placing a solid boundary far enough away from the domain of interest so that it does not interact with the flow. At present in the JHU-SPH scheme, there are two options to model the boundary-water particle interaction.

(i) Repulsive force boundary condition

One method to model solid boundaries is by means of a repulsive force with a singularity as the distance between an approaching water particle and the wall tends to zero (Monaghan and Kos 1999). The force experienced by a water particle, **f**, acts normal to the wall and is given by

$$\mathbf{f} = \mathbf{n} R(y) P(x) \varepsilon(z, u_\perp), \qquad (2.16)$$

where **n** is the normal of the solid wall. The distance y is the perpendicular distance of the particle from the wall, while x is the projection of interpolation location $\mathbf{x}_i$ onto the chord joining the two adjacent boundary particles. The repulsion function, $R(y)$, is evaluated in terms of the normalized distance from the wall, $q = y/2h$, as

$$R(y) = A \frac{1}{\sqrt{q}} (1-q), \qquad (2.17)$$

where the coefficient A is

$$A = \frac{1}{h} 0.01 c_i^2. \qquad (2.18)$$

The function $P(x)$ is chosen so that a water particle experiences a constant repulsive force as it travels parallel to the wall

$$P(x) = \frac{1}{2}\left(1 + \cos\left(\frac{2\pi x}{\Delta b}\right)\right), \qquad (2.19)$$

where Δb is the distance between any two adjacent boundary particles. Finally, the function $\varepsilon(z, u_\perp)$ is a modification to Monaghan and Kos's original suggestion and adjusts the magnitude of the force according to the local water depth and velocity of the water particle normal to the boundary

$$\varepsilon(z, u_\perp) = \varepsilon(z) + \varepsilon(u_\perp), \qquad (2.20)$$

where

$$\varepsilon(z) = \begin{cases} 0.02 & z \geq 0 \\ |z/h_o| + 0.02 & 0 > z \geq -h_o \\ 1 & |z/h_o| > 1, \end{cases} \qquad (2.21)$$

and

$$\varepsilon(u_\perp) = \begin{cases} 0 & u_\perp > 0 \\ |20 u_\perp|/c_o & |20 u_\perp| < c_o \\ 1 & |20 u_\perp| > c_o. \end{cases} \qquad (2.22)$$

In Eqs. (2.20)–(2.22), z is the elevation above the local still-water level h_o, $u_\perp = (\mathbf{u}_{WaterParticle} - \mathbf{u}_{BoundaryParticle}) \cdot \mathbf{n}$ and $c_o = \sqrt{B\gamma/\rho_o}$.

(ii) Double-Particle boundary

This technique uses conventional SPH particles that remain part of the prescribed boundary. While the position and velocity of these particles is known *a priori*, the density of these particles is predicted by Eqs. (2.6a) or (2.10a). This density is then inserted into the equation of state (2.5) to give the pressure that a boundary particle exerts on a fluid particle when evaluating the momentum equations (2.6a) or (2.10a). This type of boundary condition has proved robust for three-dimensional SPH simulations (Gómez-Gesteira and Dalrymple 2004). However, when this boundary condition was used for three-dimensional simulations in Sec. 3.2, it was found to have too much wall friction which became critical near the shoreline.

3. Numerical Test Cases

The test cases carried out here are chosen from those offered for this Tsunami workshop. The other test cases involved the modeling of large domains, which would require far more descretization than feasible for our present model. This highlights one of the current limitations of SPH in that its application to huge domains remains inappropriate and beyond practical use. In all cases, the governing equations were advanced using a second-order predictor-corrector time integration scheme.

3.1. *Test 3 — Tsunami Generation and Runup Due to a Two-Dimensional Landslide*

In this test, a Gaussian-shaped landslide, initially at the shoreline moves down a sloping bed generating a free-surface displacement. An analytical solution for the linear shallow water equations is provided by Liu *et al.* (2003), so that the objective is to compare numerical results from our SPH scheme with the solution. In dimensional form, the seafloor is described by

$$h(x,t) = H(x) + h_o(x,t), \qquad (3.1)$$

where

$$H(x) = \tan \beta, \qquad (3.2a)$$

$$h_o(x,t) = \delta \exp\left[-\left(2\sqrt{\frac{x\mu^2}{\delta \tan \beta}} - \sqrt{\frac{g}{\delta}}\mu t^2\right)^2\right], \qquad (3.2b)$$

where δ is the maximum slide thickness, μ is the thickness to slide length ratio, β is the beach slope, and x is the distance from the still-water level. For the two test cases (A and B) with different bed slopes, domain lengths and landslide height-to-length ratios, numerical results will only be presented for one case.

Test A involves a domain of length 10km, maximum depth 997m where the analytical solution predicted a free-surface displacement of 0.07m. This ratio between the domain depth and free-surface displacement necessitates a prohibitively large number of equi-sized particles both in the vertical and horizontal directions. To the authors' knowledge, no SPH scheme for free-surfaces that includes variable resolution has been developed which would make the simulation of such a problem quicker and more efficient. Therefore, the JHU-SPH code was not run for this case. Test case B, however, involves a problem domain length of 130m and depth 13m, respectively. According to the analytical solution, the movement of this landslide produces a free-surface displacement of ~1.6m for which the application of SPH is straightforward.

Figure 1 shows a snapshot of the simulation with 34466 particles ($\Delta x = 0.196$m), while in Fig. 2, four plots show the comparison at different times (non-dimensionalized by $\mu^{-1}\sqrt{\delta/g}$) of the free-surface given by SPH and the linear analytical solution. The free-surface for the SPH scheme was estimated by dropping a density-depth gage from above the free-surface every metre. The density was evaluated as

$$\rho(\mathbf{x}) = \sum_j \rho_j W(\mathbf{x} - \mathbf{x}_j) V_j. \qquad (3.3)$$

The free-surface was defined as the location at which $\rho(\mathbf{x})$ exceeds $\frac{1}{2}\rho_o = 500 \text{ kg/m}^3$.

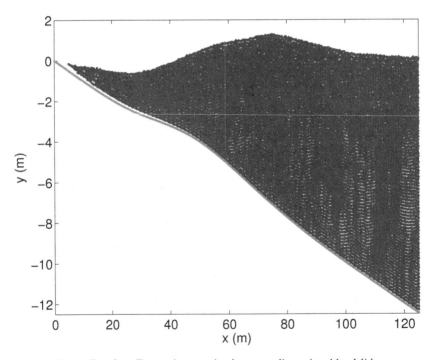

Fig. 1. Test 3 — Tsunami generation by a two-dimensional landslide.

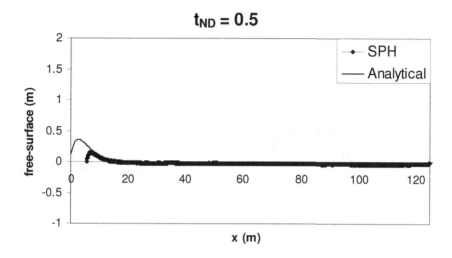

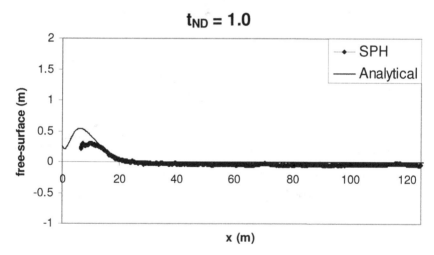

Fig. 2. (a and b) Test 3 — Tsunami generation by a two-dimensional landslide, comparison with analytical solution of linear shallow water equations.

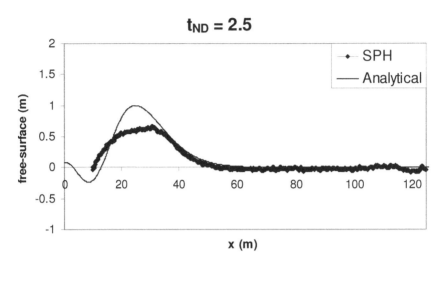

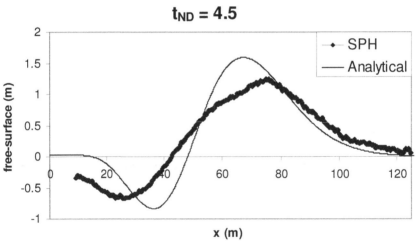

Fig. 2. (c and d) Test 3 — Tsunami generation by a two-dimensional landslide, comparison with analytical solution of linear shallow water equations.

In plots (a and b) we can see that as the solution initially develops, the discrepancy is greatest near the shoreline where the Lagrangian nature of the solution gives a different answer from that predicted by the analytical solution which uses a fixed shoreline position with the movement of the bed. As the solution develops in plots (c and d), we can see that SPH differs considerably from the linear analytical solution in that the left-to-right traveling crest is lower and further advanced for SPH. Similarly, the trough in SPH is nearer the shoreline than predicted analytically. These results are in qualitative agreement with results from nonlinear depth-averaged approaches (e.g. Liu *et al.* 2003). Thus, it is very encouraging that SPH captures inherently the nonlinearity of the flow in this test case even when using a quadratic kernel.

3.2. Test 4 — Tsunami Generation and Runup Due to a 3-Dimensional Landslide

This problem requires the modeling of a sliding mass down a 1:2 plane beach slope and compares the predictions from the JHU-SPH scheme with laboratory data. Large-scale experiments were conducted in a wave tank with a length 104m, width 3.7m, depth 4.6m and with a plane slope (1:2) located at one end of the tank (Raichlen and Synolakis 2003). A solid wedge was used to model the landslide. The triangular face has the following dimensions: a horizontal length of $b = 91$cm, a vertical face $a = 46$cm high and a width of $w = 61$cm. The wedge was instrumented with an accelerometer to define accurately the acceleration-time history and a position indicator to independently determine the velocity- and position-time histories.

In our simulations, we moved the wedge according to the motion of the wedge as measured experimentally. We conducted both two and three-dimensional simulations of the experiment.

3.2.1. *Two-Dimensional Simulation*

(i) Wedge initially partially submerged – Run 30

Figure 3 shows a snapshot of the particle positions. One can see that following the passage of the wedge down the slope, a large volume of water has been drawn down in generating the offshore propagating wave. This is confirmed in Fig. 4 which displays a comparison of the free-surface data at a wave gage located 1.83m offshore from the still-water line. The 2-D SPH-LES scheme vastly overpredicts both the initial wave and the following drawdown of the free-surface. This simulation was performed using 8600 particles where $\Delta x = 0.09$m taking 3 hours of cpu time. A convergence study showed no change in the predicted numerical behavior allowing us to conclude that the cause of such a large discrepancy is that important three-dimensional effects of the drawdown and violent splash collisions are neglected.

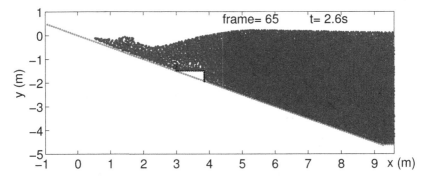

Fig. 3. Test 4 — Tsunami generation by a three-dimensional landslide, run 30 (initially partially submerged): 2-D SPH-SPS simulation snapshot.

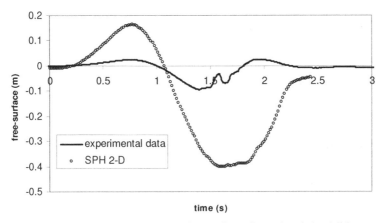

Fig. 4. Test 4 — Tsunami generation by a three-dimensional landslide, run 30: Comparison of 2-D SPH-SPS with wave gage 1 data.

(ii) Wedge initially completely submerged – Run 32

When the wedge is placed just under the water surface so that it is only just completely submerged, the simulation exhibits the same behavior as for run 30 above. Figure 5 shows the comparison of the water surface at the same wave gage. The same large initial wave and drawdown are visible. A similar convergence study again confirmed the lack of three dimensionality.

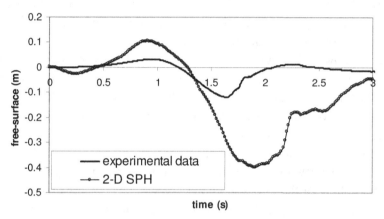

Fig. 5. Test 4 — Tsunami generation by a three-dimensional landslide, run 32: (initially partially submerged): Comparison of 2-D SPH-SPS with wave gage 1 data.

3.2.2. Three-Dimensional Simulation

The two-dimensional simulations demonstrate the necessities and requirements of a three-dimensional simulation. The extra spatial dimension brings a further demand on the computational cost of using SPH for flows to which as a numerical technique it is otherwise ideally suited.

Our experience of using the double-particle boundary described in Sec. 2.3 was that this boundary condition caused too much wall friction. This had the undesirable effect that in the wake behind the sliding block, the particles would not occupy the space left behind by the block.

Thus, a 3-D version of the adjusted Monaghan boundary condition was used. The simulation was performed on a single processor machine with 512MB RAM with 62377 particles taking approximately 80 hours of cpu. The computer resources available limited the discretization to $\Delta x = \Delta y = \Delta z = 0.1$m. Figures 6–8 display three snapshots from the simulation. Each snapshot shows four particle plots from different views: a front view (with the block moving towards the viewer), a side view, a 3-D view and a top view looking down on the domain. In Fig. 6, we see the simulation just after the block has entered the water: from the side view, there is a very clear surface wave being generated by the initial push of the block. In Fig. 7, the block has continued to slide down the slope with the initial wave continuing to propagate offshore; from the top view, we can see that the water from the either side of the wedge is now beginning to flow and occupy the space vacated by the moving block. In Fig. 8, the water flowing to occupy the space behind the block has now collided with itself generating a "rooster tail" or violent upward splash event (see side view and front view) as observed in the experiments. There is then a wave that propagates out from the splash-up which is not shown here.

If we identify the free-surface position at the wave-gage located 1.83m offshore of the initial shoreline, we see the comparison of the 3-D SPH results with the 2-D and experimental data in Fig. 9. There is a clear improvement moving from 2-D to 3-D in that the size of the wave and drawdown generated is smaller. However, there is still a very large discrepancy between SPH and experimental data.

This is for several reasons. Primarily, the resolution of $\Delta x = \Delta y = \Delta z = 0.1$m is clearly insufficient to capture free-surface displacements and flow features whose scale is on the order of 0.02m. Other contributing factors include the size of the domain that has been dictated by computational costs. Furthermore, at present, no method in SPH exists to model offshore non-reflecting boundaries properly for free-surface flows. Hence, here, periodic lateral boundaries were used in the longshore direction, while a wall was placed offshore. Neither of these boundaries was at a sufficient distance from the moving block, (i.e. if computationally possible, we would have placed the offshore wall much further from the moving block). Other sources of error include the non-dynamic sub-particle scale turbulence model and boundary conditions.

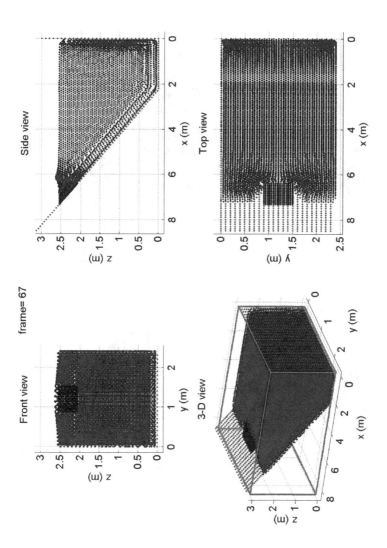

Fig. 6. Test 4 — Tsunami generation by a three-dimensional subaerial landslide, run 30: Monaghan boundary condition, 3-D SPH simulation, snapshot 1.

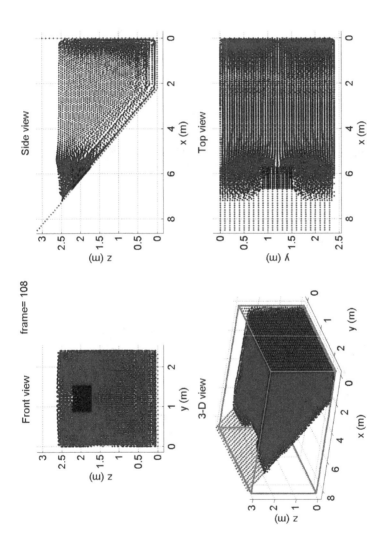

Fig. 7. Test 4 — Tsunami generation by a three-dimensional subaerial landslide, run 30: Monaghan boundary condition, 3-D SPH simulation, snapshot 2.

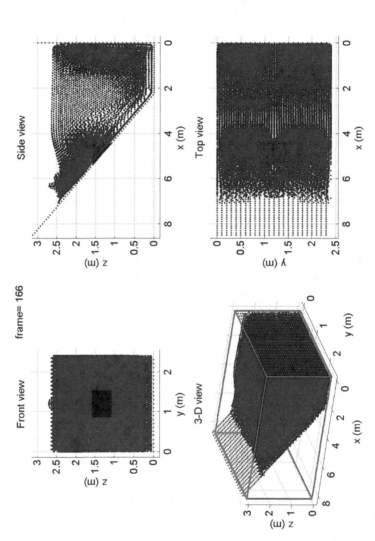

Fig. 8. Test 4 — Tsunami generation by a three-dimensional subaerial landslide, run 30: Monaghan boundary condition, 3-D SPH simulation, snapshot 3.

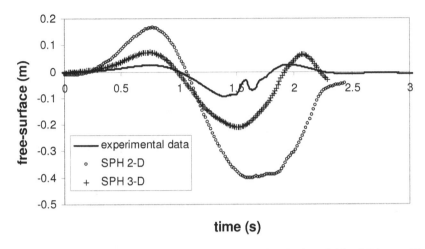

Fig. 9. Test 4 — Tsunami generation by a three-dimensional sub-aerial landslide, run 32: Comparison of 3-D and 2-D SPH-SPS with wave gage 1 data.

Overall however, the results from this case clearly demonstrate one of the essential drawbacks of SPH. The resolution needed to perform an accurate simulation of this test case must be on the order of O (Δx, Δy, Δz) ≈ 0.005m. For the dimensions of the current case, this would mean ~ 32×10^7 particles, which can only be accomplished using high-performance parallel computing, this is unavailable to the authors at the present time.

4. Conclusions and Future Work

Smoothed particle hydrodynamics has been presented as method that can be applied to free-surface hydrodynamics. The method can be used straightforwardly for two-dimensional problems where the ratio of the free-surface displacement to the depth and length of the domain is not prohibitively small. For the 2-D landslide generated tsunami, the SPH-LES method captures inherently nonlinear effects.

Looking to the future, clearly, due to its high computational cost, improving computer power and resources will make SPH an increasingly viable method for simulating tsunami flows. In the meantime, SPH would be best used in conjunction with other models that predict the far-field flow which are then linked to SPH which models the complicated 3-D fluid-structure interaction. Hence, of particular engineering merit will be the patching together of different numerical models with SPH whereby flow information is passed between each model. At JHU, work is already underway investigating how SPH can be linked to shallow water and Boussinesq-type models (FUNWAVE from the University of Delaware) that will predict the far-flow field utilizing SPH to focus on localised flow features. Boussinesq-type models are now a well established technique for predicting the propagation of dispersive and nonlinear waves in shallow water. Recent advances (e.g. Madsen *et al.* 2003) have greatly extended the accuracy of the linear dispersion characteristics and nonlinear properties of Boussinesq-type models. In the depth-integrated approach, the modeling of wave breaking itself has been difficult requiring approximate submodels such as modifications of Svendsen's (1984) decay of a turbulent bore. SPH will clearly be able to handle the turbulent flow field with the propagation of the breaking wave.

Following tsunami propagation into shallow water, SPH will also be ideally suited to examining the three-dimensional flow fields of tsunami-interaction with coastal structures and inundation of dry land. This will allow modelers to address the important issues of tsunami-structure interaction and modeling debris flow following the passage of the tsunami wave.

A natural consequence of being able to model tsunami-structure interaction is that we also envisage using SPH to model the rupture of safety critical buildings such as oil storage tanks. In contrast to other formulations of fluids flow and intense wave breaking, SPH facilitates easily the modeling of multi-fluid flow enabling us to examine the transport and dispersal of contaminants and pollutants following such catastrophic failures.

Acknowledgment

This work was supported by the Office of Naval Research (ONR).

References

1. Batchelor, Sir. G. K., *An Introduction to Fluid Dynamics* (Cambridge University Press, 1967) pp. 635.
2. Blin, L., A. Hadjadj and L. Vervisch, Large eddy simulation of turbulent flows in reversing systems, *Selected Proceedings of the 1st French Seminar on Turbulence and Space Launchers*, CNES-Paris, 13–14 June, 2002, eds. P. Vuillermoz, P. Comte and M. Lesieur, 2002.
3. Briggs, M. J., C. E. Synolakis and G. S. Harkins, Tsunami runup on a conical island, *Proc. Waves — Physical and Numerical Modeling*, University of British Columbia, Vancouver, Canada, 446–455, 1994.
4. Christensen, E. D. and R. Deigaard, Large eddy simulation of breaking waves, *Coastal Eng.* **42**, 53–86, 2001.
5. Dalrymple, R. A. and O. Knio, SPH modeling of water waves, *Proc. Coastal Dynamics 2001*, ASCE, 779–787, Lund, Sweden, 2001.
6. Dalrymple, R. A., O. Knio, D. T. Cox, M. Gesteira, and S. Zou, Using a Lagrangian particle method for deck overtopping, *Proc. Ocean Wave Measurement and Analysis*, ASCE, 1082–1091, 2001.
7. Gingold, R. A. and J. J. Monaghan, Smoothed particle hydrodynamics: theory and application to non-spherical stars, *Mon. Not. R. Astron. Soc.*, **181**, 375–389, 1977.
8. Gómez-Gesteira, M. and R. A. Dalrymple, Using a 3D SPH method for wave impact on a tall structure, *J. Waterways, Port, Coastal, and Ocean Engineering*, ASCE, **130**(2), 63–69, 2004.
9. Gotoh, H., T. Sakai and M. Hayashi, Lagrangian model of drift-timbers induced flood by using moving particle semi-implicit method, *J. Hydroscience and Hyd. Eng.* **20**(1), 95–102, 2002.
10. Gotoh, H., T. Shibihara and T. Sakai, Sub-particle-scale model for the MPS method — Lagrangian flow model for hydraulic engineering, *Computational Fluid Dynamics J.*, **9**(4), 339–347, 2001.
11. Gotoh, H., S. Shao and T. Memita, SPH-LES model for numerical investigation of wave interaction with partially immersed breakwater, *Coastal Engineering J.*, **46**(1), 39–63, 2004.
12. Grilli, S. T., S. Vogelmann and P. Watts, Development of a 3D Numerical Wave Tank for modeling tsunami generation by underwater landslides, *Eng. Analysis Boundary Elemt.* **26**(4), 301–313, 2002.
13. Johnson, G. R. and S. R. Beissel, Normalized smoothing functions for SPH impact computations, *Int. J. Num. Meth. Eng.* **39**, 2725–2741, 1996.
14. Koshizuka, S., A. Nobe and Y. Oka, Numerical analysis of breaking waves using the moving particle semi-implicit method, *Int. J. Numer. Meth. Fluids*, **26**, 751–769, 1998.
15. Li, S. and W. K. Liu, Meshfree and particle methods and their applications, *Appl. Mech. Rev.* **55**(1), 1–34, 2002.

16. Li, Y. and F. Raichlen, Energy balance model for breaking solitary wave runup, *J. Waterways, Port, Coastal and Ocean Engineering*, 47–59, 2003.
17. Lin, P., K.-A. Chang and P. L.-F. Liu, Runup and Rundown of Solitary Waves on Sloping Beaches, *Journal of Waterways, Port, Coastal and Ocean Engineering*, **125**(5), 247–255, 1999.
18. Liu, P. L.-F., P. Lynett and K. Synolakis, Analytical solutions for forced long waves on a sloping beach, *Journal of Fluid Mechanics*, **478**, 101–109, 2003.
19. Lo, E. Y. M. and S. Shao, Simulation of near-shore solitary wave mechanics by an incompressible SPH method, *Applied Ocean Research*, **24**, 275–286, 2002.
20. Lucy, L. B., A numerical approach to the testing of fusion process, *Astronomical J.*, **88**, 1013–1024, 1977.
21. Lynett, P., T.-R. Wu and P. L.-F. Liu, Modeling Wave Runup with Depth-Integrated Equations, *Coastal Engineering*, **46**(2), 89–107, 2002.
22. Madsen, P. A., H. B. Bingham and H. A. Schaffer, Boussinesq-type formulations for fully nonlinear and extremely dispersive water waves: derivation and analysis, *Proc. Roy. Soc. London*, Series A, **459**(2033), 1075–1104, 2003.
23. McGuire, B., *Apocalypse: a Natural History of Global Disasters*, Cassell, London, 1999.
24. Monaghan, J. J., Smoothed particle hydrodynamics, *Annual Review of Astronomy and Astrophysics*, **30**, 543–574m, 1992.
25. Monaghan, J. J., Simulating free surface flows with SPH, *J. Computational Physics*, **110**, 399–406, 1994.
26. Meneveau, C. and J. Katz., Scale-invariance and turbulence models for large-eddy simulation, *Annual Rev. Fluid Mechanics*, **32**, 1–32, 2000.
27. Monaghan, J. J., SPH without a tensile instability, *J. Computational Physics*, **159**, 290–311, 2000.
28. Monaghan, J. J. and A. Kos, Solitary waves on a Cretan beach, *J. Waterway, Port, Coastal, and Ocean Eng*, ASCE, **125**, 3, 1999.
29. Morris, J., P. Fox and Y. Zhu, Modeling low Reynolds number incompressible flows using SPH, *Journal of Computational Physics*, **136**, 214–226, 1997.
30. Panizzo, A. and R. A. Dalrymple, SPH Modelling of Underwater Landslide Generated Waves, *Proc. 29th Intl. Conference on Coastal Engineering*, Lisbon, World Scientific Publishing Co., 1147–1159, 2004.
31. Raichlen, F. and C. E. Synolakis, Run-up from three dimensional sliding mass, *Proceedings of the Long Wave Symposium 2003*, (Briggs, M, Coutitas, Ch.) XXX IAHR Congress Proceedings, 247–256, 2003.
32. Randles, P. W. and L. D. Libersky, Smoothed Particle Hydrodynamics: Some recent improvements and applications, *Comput. Methods Appl. Mech. Eng.*, **139**, 375–408, 1996.
33. Shankar, N. J. and M. P. R. Jayaratne, Wave run-up and overtopping on smooth and rough slopes of coastal structures, *Ocean Engineering*, **30**(2), 153–295, 2003.

34. Shao S. D., and H. Gotoh, Simulating coupled motion of progressive wave and floating curtain wall by SPH-LES model, *Coastal Engineering Journal*, **46**(2), 171–202, 2004.
35. Smagorinsky, J., General circulation experiments with the primitive equations. I. The basic experiment, *Mon. Weather Rev.*, **91**, 99–164, 1963.
36. Svendsen, I., Wave heights and set-up in a surf zone, *Coastal Engineering*, **8**, 303–329, 1984.
37. Toro, E. F., M. Olim and K. Takayama, Unusual increase in tsunami wave amplitude at the Okushiri island: Mach reflection of shallow water waves, *Proc. 22nd Int. Symp. on Shock Waves*, Imperial College, London, UK, July 18–23 1999.
38. Tulin, M. P. and M. Landrini, Breaking waves in the ocean and around Ships, *Proc. 23rd ONR Symposium on Naval Hydrodynamics*, Val de Reuil, France, 2000.
39. Yeh, H. and M. Petroff, Bore in a box experiment, http://engr.smu.waves.edu/solid.html, 2003.
40. Yoshizawa, A., Statistical theory for compressible turbulent shear flows, with the application to subgrid modeling, *Phys. Fluids A*, **29**, 2152–2164, 1986.

CHAPTER 4

A LARGE EDDY SIMULATION MODEL FOR TSUNAMI AND RUNUP GENERATED BY LANDSLIDES

Tso-Ren Wu[*] and Philip L.-F. Liu

School of Civil and Environmental Engineering, Cornell University
Ithaca, NY, 14850, U.S.A.
E-mail: [*]tw36@cornell.edu

A numerical model is developed to simulate landslide-generated tsunamis. This numerical model is based on the Large Eddy Simulation (LES) approach, solving three-dimensional filtered Navier-Stokes equations. While the Smagorinsky subgrid scale (SGS) model is employed for turbulence closure, the Volume of Fluid (VOF) method is used to track the free surface and shoreline movements. A numerical algorithm describing the movements of a solid landslide is also developed. The numerical results are compared with laboratory experimental data (The benchmark problem 4). A very good agreement is shown for the time histories of runup and generated waves. Based on the numerical results, the detailed 3D complex flow patterns, free surface and shoreline deformations are further analyzed.

1. Introduction

Tsunamis could be generated by many different geophysical phenomena, such as earthquake, landslide, and volcano eruption. Aerial and submarine landslides have been documented as a source of several destructive tsunamis. For example, in 1958 a rockslide, triggered by an 8.3 magnitude earthquake, occurred in Lituya Bay, Alaska. An estimated volume of 30.6 million m^3 of amphibole and biotite slide down into the Gilbert Inlet at the head of Lituya Bay, generating large tsunamis. The runup height on the opposite side of the slide in the Gilbert Inlet was estimated as 524 m. On November 29, 1975, a landslide was triggered by a 7.2-magnitude earthquake along the southeast coast of Hawaii. A 60 km stretch of Kilauea's coast subsided 3.5 m and moved seaward 8 m. This landslide generated a local tsunami with a maximum run-up height of 16 m at Keauhou (Cox and Morgan, 1977). More recently, the Papua New Guinea (PNG) tsunami in 1998 is thought to have been generated by a submarine landslide, triggered by an earthquake of magnitude 7 (Bardet et al., 2003; Kawata et al., 1999; Synolakis et al., 2002). The PNG tsunami killed more than 4000 people and was considered as one of most devastating tsunamis in recent history.

Modeling the landslide generated tsunamis faces many challenges, such as the complex three-dimensional interfacial flow structures between water and slide materials, the moving free surface and shoreline boundaries, and the strong local turbulence in the vicinity of the slide and near the free surface. So far most of the numerical simulation models for landslide-generated waves have been built upon the depth-integrated continuity and momentum equations (Jiang and LeBlond, 1994; Lynett et al., 2002). However, the basic assumptions employed in the depth-integrated models, such as the long wave approximation with weak vertical acceleration, limit their applications to small and smooth slides.

An alternative to the depth-integrated model is to use the Boundary Element Method (BEM) to solve the three-dimensional fully nonlinear potential flow (Grilli et al., 2002). The advantage of the BEM is its ability in solving three-dimensional flows with both moving solid boundaries and free surface. However, the potential flow assumption limits its applications to mostly non-breaking wave problems.

In order to study landslide-generated waves more precisely, it is desirable to solve the three-dimensional Navier-Stokes (N-S) equations. Theoretically, the direct numerical simulation (DNS) can be performed to resolve both large eddy motions and the smallest turbulence (Kolmogorov) scale motion. Therefore, DNS requires very fine spatial and temporal resolutions and most of DNS applications can only be applied to relatively low Reynolds number flows within a small computational domain (Kim et al., 1987). Landslide generated breaking waves are high Reynolds number flows with strong free surface deformation. With the current available computing resources, the DNS does not seem to be a feasible approach for solving this type of problems.

The alternatives to the DNS approach for computing the turbulent flow characteristics include the Reynolds Averaged Navier-Stokes (RANS) equations method and the Large Eddy Simulation (LES) method. In the RANS equations method, only the ensemble-averaged (mean) flow motion is resolved. The turbulence effects appear in the momentum equations for the mean flow and are represented by Reynolds stresses, which are modeled by an eddy viscosity model. The eddy viscosity is hypothesized as a function of the turbulence kinetic energy (k) and the turbulence dissipation rate (ε), for which balance equations are constructed semi-empirically. Therefore, this method is also referred to as the $k-\varepsilon$ turbulence model. Recently, Lin and Liu (1998a, 1998b) have successfully applied the $k-\varepsilon$ turbulence model in their studies of wave breaking and runup in the surf zone. Lin and Liu's model has been extended and applied to many different coastal engineering problems, including the wave-structure interaction [Liu *et al.* (1999, 2001)]. In the LES method, the three-dimensional turbulent motions are directly simulated and resolved down to a pre-determined scale, and the effects of smaller-scale motions are modeled by closures. In terms of the computational expense, LES lies between RANS and DNS. Comparing to DNS in solving high-Reynolds-number flows, LES avoids explicitly representing small-scale motions and therefore, the computational cost can be greatly reduced. Comparing to RANS models, because the large-scale unsteady motions are computed explicitly, LES can be expected to provide more statistical information for the turbulence flows in which large-scale unsteadiness is significant (Pope, 2001). In this paper, a LES approach will be presented for solving the landslide-generated water waves and runup.

The flow equations for LES are derived from the NS equations by applying a low-pass spatial filter. Similar to the RANS approach, a residual-stress tensor or the sub-grid-scale (SGS) Reynolds stress tensor appears in the filtered N-S equations. Thus, a closure model is also required to relate the residual-stress tensor to the filtered velocity field.

The traditional Smagorinsky model (Smagorinsky, 1963) is probably the simplest LES SGS model and has been used in several breaking wave studies (Watanabe and Saeki, 1999; Christensen and Deigaard, 2001; Lin and Li, 2003). In this study, we shall adopt the traditional LES approach with the Smagorinsky SGS model.

The landslide-generated wave problem is further complicated due to the appearance of a free surface. The location of the free surface is a part of solution to be determined. The boundary fitting method (Lin and Li, 2002) is one of many ways to locate the free surface, in which the free surface is treated as a sharp interface. The boundary fitted grids are employed and the computational domain must be re-meshed each time as the free surface moves. This method gives highly accurate results of tracking the free surface. However, due to the difficulty of fitting the free surface associated with breaking waves, where air bubbles are entrained into water body, this method is not practical for the present problem.

Other available methods for tracking free surface movements include the Marker-and-Cell (MAC) method (Harlow and Welch, 1965) and the Volume-of-Fluid (VOF) (Hirt and Nichols, 1981) method. Different from the boundary fitting method, these two schemes do not define the interface as a sharp interface and computations are performed on a stationary grid. In the MAC method, massless particles are introduced initially on the free surface. The free surface is determined by following the motions of these massless particles. The MAC method is appealing because they can accurately treat the complex phenomena such as wave breaking. However, the computing effort of the MAC scheme could be daunting, especially in three dimensions because in addition to solving the equations governing the fluid flow, one has to trace the motions of a large number of marker particles.

In the VOF method, along with the conservation equations for mass and momentum, one only needs to solve one additional equation for the fraction factor f assigned for each numerical cell, where f is defined as follows: $f = 1$ in a water cell and $f = 0$ in an air cell, and $0 < f < 1$ indicates a free surface cell. Beside this additional transport equation, a free surface reconstruction algorithm is used to determine the orientation of the free surface. The VOF method has been successfully implemented for the simulation of various breaking wave related problems [e.g., Lin and Liu (1998a, 1998b)]. In their studies, the interface between air and water was reconstructed by either a horizontal or a vertical line. This piecewise constant interface reconstruction scheme is relatively crude. Rider and Kothe (1998) provided a second-order reconstruction scheme, in which the interface is approximated as piecewise linear. In this paper, we will use the VOF method and adopt the piecewise linear interface reconstruction scheme in a LES model.

In addition to the treatment of free surface, modeling the landslide-generated waves faces another difficulty: the treatment of moving solid boundaries (i.e., landslides). One possible method is to use curvilinear boundary-fitted coordinates. However, for a problem with moving or deforming boundaries the boundary-fitted grids must be either regenerated or deformed as the boundary geometry changes, adding considerable complexity to the computations. Another approach is the immersed boundary method (IBM). This method introduces a body-force field so that a desired velocity distribution can be assigned over a solid boundary (e.g., Mohd-Yusof, 1997; Goldstein *et al.*, 1993). This method allows the imposition of the boundary conditions on a given surface which does not coincide with the computational grids. The IBM has been successfully applied to the internal combustion (IC) piston simulations (Fadlun *et al.*, 2000; Verzicco *et al.*, 2000). The IBM has the advantage of providing highly accurate results for moving boundary problems. However, since this method depends on the interpolated velocity on the boundary to evaluate the immersed body-force, it requires fine resolution on the solid boundary.

Heinrich (1992) developed a partial cell method, in which a source function is added to the continuity equation to represent the solid boundary movement. He applied this method to simulate waves generated by a two-dimensional landslide and obtained very good numerical results for water surface profiles as compared to the laboratory data. This moving solid algorithm does not require any interpolation scheme and is independent from the grid system. Thus, it is extended to three-dimensional problems in the current study.

The present numerical examples concern the solid landslide motion as well as the landslide generated breaking wave runup and rundown on a slopping beach. A subaerial and a submerged landslide cases will be addressed. The submerged landslide case is the benchmark problem #4 discussed in the Third International Workshop on Long-Wave Runup Models held at Wrigley Marine Science Center, Catalina Island, California (2004).

The numerical results will be compared with the experimental data (Raichlen and Synolakis, 2003) for time histories of free surface fluctuations. The numerical results will be used to illustrate the complex 3D flow patterns in terms of the velocity field, shoreline movement, and free surface profiles.

2. Governing Equations and Boundary Conditions

2.1. *The Navier-Stokes Equations*

The fluid motions of an incompressible Newtonian fluid can be described by the N-S equations:

$$\nabla \cdot \mathbf{u} = 0, \tag{2.1}$$

$$\frac{\partial \mathbf{u}}{\partial t} + \nabla \cdot (\mathbf{u}\mathbf{u}) = -\frac{1}{\rho_w}\nabla p + \mathbf{g} + \frac{1}{\rho_w}\nabla \cdot \tilde{\tau}, \tag{2.2}$$

where $\mathbf{u}$ represents the velocity vector, ρ_w water density, $\tilde{\tau}$ the stress tensor, $\mathbf{g}$ the gravity force vector, t time, and p pressure. The viscous stress tensor is a function of the molecular viscosity μ_w and the strain rate $\nabla \mathbf{u}$:

$$\tilde{\tau} = \mu_w \left(\nabla \mathbf{u} + \nabla^T \mathbf{u}\right). \tag{2.3}$$

The above equations can be also written in the tensor format:

$$\frac{\partial u_i}{\partial x_i} = 0, \tag{2.4}$$

$$\frac{\partial u_i}{\partial t} + \frac{\partial u_i u_j}{\partial x_j} = -\frac{1}{\rho_w}\frac{\partial p}{\partial x_i} + g_i + \frac{1}{\rho_w}\frac{\partial \tau_{ij}}{\partial x_j}, \quad (2.5)$$

where $i, j = 1, 2, 3$ for three-dimensional flows, and τ_{ij} is the molecular viscous stress tensor:

$$\tau_{ij} = \mu_w \left(\frac{\partial u_i}{\partial x_j} + \frac{\partial u_j}{\partial x_i} \right). \quad (2.6)$$

In order to solve the NS equations, the initial and boundary conditions have to be applied. Along a solid boundary, Γ_1, (Fig. 2.1) the fluid velocity must be the same as the boundary velocity, u_{si}. Thus the boundary condition is written as:

$$u_i = u_{si}. \quad (2.7)$$

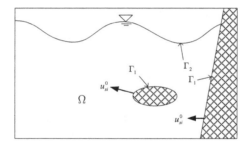

Fig. 2.1. A sketch of the flow domain and boundaries. The gray parts indicate the solid material.

Another type of boundary is the free surface, Γ_2, which is the interface between air and water. Along the free surface, both dynamic and kinematic boundary conditions are needed. Without considering the surface tension, the normal and tangential stress components must be continuous, which can be expressed as:

$$\left.\begin{array}{l}-p+2\mu_w\dfrac{\partial u_n}{\partial n}=S_n\\[2mm]\mu_w\left(\dfrac{\partial u_{T_k}}{\partial n}+\dfrac{\partial u_n}{\partial T_k}\right)=S_{T_k}\end{array}\right\},\quad \text{on } \Gamma_2 \qquad (2.8)$$

where the subscripts n and T_k denote the outward normal direction and two tangential directions ($k=1, 2$), respectively, on the free surface. S_n and S_{T_k} are the normal and tangential stress components specified by the air flow on the free surface. The kinematic boundary condition requires that the free surface be a sharp interface separating the two fluids so that no flow is allowed to go through it. The mathematical expression of the kinematic boundary can be derived from the equation that describes the free surface. If the free surface is expressed as $J(\mathbf{x},t)=0$, the total derivative of $J(\mathbf{x},t)$ must be zero on the surface:

$$\left.\frac{DJ(\mathbf{x},t)}{Dt}=0\right|_{\text{on } J(\mathbf{x},t)=0 \text{ or on } \Gamma_2} \qquad (2.9)$$

or

$$\frac{\partial J}{\partial t}+\mathbf{u}\cdot\nabla J=0 \qquad \text{on } J(\mathbf{x},t)=0, \text{ or on } \Gamma_2. \qquad (2.10)$$

For the initial conditions, the whole flow field has to be prescribed:

$$u_i(\mathbf{x},0)=u_i^0(\mathbf{x}), \qquad (2.11)$$

and the initial value of J can be represented as:

$$J(\mathbf{x},0)=J^0(\mathbf{x}). \qquad (2.12)$$

Therefore, the initial pressure, p^0, can be calculated from Eq. (2.5) based on the initial velocity field, $u_i^0(\mathbf{x})$. Finally, the initial velocity and pressure fields must satisfy the following conditions (see Fig. 2.1 for a sketch of the fluid domain and boundaries):

$$\frac{\partial u_i^0}{\partial x_i} = 0, \qquad \text{in } \Omega$$

$$\frac{\partial u_i^0}{\partial t} + \frac{\partial u_i^0 u_j^0}{\partial x_j} = -\frac{1}{\rho_w}\frac{\partial p^0}{\partial x_i} + g_i + \frac{1}{\rho_w}\frac{\partial \tau_{ij}^0}{\partial x_j}, \qquad \text{in } \Omega$$

$$u_i^0 = u_{si}^0, \qquad \text{on } \Gamma_1$$

$$-p^0 + 2\mu_w \frac{\partial u_n^0}{\partial n} = S_n^0, \qquad \text{on } \Gamma_2 \qquad (2.13)$$

$$\mu_w \left(\frac{\partial u_{T_k}^0}{\partial n} + \frac{\partial u_n^0}{\partial T_k} \right) = S_{T_k}^0, \qquad \text{on } \Gamma_2$$

$$\frac{\partial J^0}{\partial t} + u_i^0 \frac{\partial J^0}{\partial x_i} = 0, \qquad \text{on } \Gamma_2.$$

2.2. Large Eddy Simulation

The flow motions generated by a landslide movement can be described by the NS equations with appropriate initial and boundary conditions as shown in the previous section. However, the size of the physical domain for this type of problem is usually very large and makes the DNS nearly impossible. In this section we briefly summarize the LES model (Deardorff, 1970), which solves the large-scale eddy motions and models the small-scale turbulent fluctuations. The Smagorinsky model will be adopted to close the residual stress in the filtered NS equation.

2.2.1. Filtering

In the LES, a low-pass filter is performed so that the filtered velocity field can be resolved on a relatively coarse grid. A filtered variable (denoted by an overbar) is defined by (Leonard, 1974):

$$\overline{\phi}(x) = \int \phi(x')G(x,x')dx', \qquad (2.14)$$

where G is the filter function, which can be a Gaussian, a box (a local average), or a cutoff (used in the spectrum method) filter. Each filter has a specified filter width Δ associated with it. Roughly speaking eddies of size smaller than Δ are small eddies and need to be modeled. The specified filter function G satisfies the normalization condition:

$$\int G(x,x')dx' = 1. \qquad (2.15)$$

In this paper, since the finite-volume method (FVM) is used, the FVM provides a natural filtering operation:

$$\bar{\phi}(x) = \frac{1}{V}\int_V \phi(x')dx', \, x' \in V, \qquad (2.16)$$

where V is the volume of a computational cell. Therefore, the filter function G here is specified as:

$$G(x,x') = \begin{cases} 1/V & \text{for } x' \in V \\ 0 & \text{otherwise} \end{cases}. \qquad (2.17)$$

2.2.2. Filtered Navier-Stokes Equations

The filtered N-S equations can be derived by applying the filter function to the N-S equations. The filtered continuity and momentum equations are:

$$\overline{\left(\frac{\partial u_i}{\partial x_i}\right)} = \frac{\partial \bar{u}_i}{\partial x_i} = 0, \qquad (2.18)$$

$$\frac{\partial(\bar{u}_i)}{\partial t} + \frac{\partial(\overline{u_i u_j})}{\partial x_j} = -\frac{1}{\rho_w}\frac{\partial \bar{p}}{\partial x_i} + g_i \\ + \frac{1}{\rho_w}\frac{\partial}{\partial x_j}\left[\mu_w\left(\frac{\partial \bar{u}_i}{\partial x_j} + \frac{\partial \bar{u}_j}{\partial x_i}\right)\right], \qquad (2.19)$$

where $\bar{u}$ and $\bar{p}$ are filtered velocity and pressure fields, respectively. It is important to reiterate here that, the filtered product, $\overline{u_i u_j}$, is different from the product of the filtered velocity, $\bar{u}_i \bar{u}_j$:

$$\overline{u_i u_j} \neq \bar{u}_i \bar{u}_j. \tag{2.20}$$

The difference is also called the residual-stress tensor:

$$\tau_{ij}^R \equiv \overline{u_i u_j} - \bar{u}_i \bar{u}_j. \tag{2.21}$$

In LES, τ_{ij}^R is also called the SGS Reynolds stress. The residual kinetic energy is

$$k_r \equiv \frac{1}{2}\tau_{ii}^R, \tag{2.22}$$

and the anisotropic residual-stress tensor is defined by

$$\tau_{ij}^r \equiv \tau_{ij}^R - \frac{2}{3}k_r \delta_{ij}. \tag{2.23}$$

The isotropic residual stress is included in the modified filtered pressure

$$\bar{\bar{p}} \equiv \bar{p} + \frac{2}{3}k_r. \tag{2.24}$$

Substituting (2.23) and (2.24) into (2.19), the filtered momentum equation can be rewritten as:

$$\frac{\bar{D}(\bar{u}_i)}{\bar{D}t} = -\frac{1}{\rho_w}\frac{\partial \bar{\bar{p}}}{\partial x_i} + g_i \\ + \frac{1}{\rho_w}\frac{\partial}{\partial x_j}\left[\mu_w\left(\frac{\partial \bar{u}_i}{\partial x_j} + \frac{\partial \bar{u}_j}{\partial x_i}\right)\right] - \frac{1}{\rho_w}\frac{\partial \tau_{ij}^r}{\partial x_j}, \tag{2.25}$$

where the substantial derivative based on the filtered velocity is:

$$\frac{\bar{D}}{Dt} \equiv \frac{\partial}{\partial t} + \bar{\mathbf{u}} \cdot \nabla. \qquad (2.26)$$

2.2.3. Smagorinsky Model

In (2.25), the SGS Reynolds stress contains the local average of the small scale field, therefore the SGS model should be based on the local velocity field. The most commonly used SGS model is the Smagorinsky model (Smagorinsky, 1963). It is essentially a linear eddy viscosity model:

$$\tau_{ij}^r = -\nu_t \left(\frac{\partial \bar{u}_i}{\partial x_j} + \frac{\partial \bar{u}_j}{\partial x_i} \right) = -2\nu_t \bar{S}_{ij}. \qquad (2.27)$$

The Smagorinsky model relates the residual stress to the filtered strain rate. The coefficient $\nu_t(\mathbf{x}, t)$ is the subgrid-scale eddy viscosity of the residual motions. Based on the dimensional analysis, the subgrid-scale eddy viscosity is then modeled as:

$$\begin{aligned} \nu_t &= \ell_s^2 \bar{\mathbf{S}} \\ &= (C_S \Delta)^2 \bar{\mathbf{S}}, \end{aligned} \qquad (2.28)$$

where ℓ_s is the Smagorinsky length scale, which is a product of the Smagorinsky coefficient C_S and the filter width Δ; $\bar{\mathbf{S}}$ is the characteristic filtered rate of strain:

$$\bar{\mathbf{S}} \equiv \left(2 \bar{S}_{ij} \bar{S}_{ij} \right)^{1/2}. \qquad (2.29)$$

In the finite volume discretization, Δ can be represented by the order of magnitude of grid size and is defined as:

$$\Delta = \left(dx_1 \times dx_2 \times dx_3 \right)^{1/3}, \qquad (2.30)$$

where dx_1, dx_2, and dx_3 are the three components of the grid lengths. Under the isotropic turbulence condition, the Smagorinsky coefficient $C_S \approx 0.2$. However, in general, C_S is not a constant; its value varies from 0.1 to 0.2 in different flows. Following the suggestion made by previous work for wave-structure interactions (Lin and Li 2003), the present simulations have used a value of 0.15. We remark here that existing literatures have also suggested that the Smagorinsky SGS model needs to be modified in the region close to the free surface (i.e., Shen and Yue 2001). A mixed dynamic SGS model has been developed for complex flows (Dommermuth *et al.*, 2002; Hendrikson *et al.*, 2003). In present simulations we have decided to use the Smagorinsky SGS model for its simplicity (C_s is the only empirical coefficient used in the entire simulation model). The dynamic SGS model approach will be explored in the future work.

The momentum equations for the LES with the Smagorinsky SGS model can be expressed as follows:

$$\frac{\partial \overline{u}_i}{\partial t} + \frac{\partial \overline{u}_i \overline{u}_j}{\partial x_j} = -\frac{1}{\rho_w}\frac{\partial \overline{p}}{\partial x_i} + g_i \\ + \frac{1}{\rho_w}\frac{\partial}{\partial x_j}\left[\mu_{w,eff}\left(\frac{\partial \overline{u}_i}{\partial x_j} + \frac{\partial \overline{u}_j}{\partial x_i}\right)\right], \quad (2.31)$$

where $\mu_{w,eff} = \mu_w + \mu_t$ is the sum of molecular and eddy viscosities.

2.2.4. *Near-Wall Treatment*

In the near-wall viscous sub-layer region, the largest local turbulent eddy sizes are limited by the viscous scales. The well resolved LES requires grids nearly as fine as those used in the DNS. This restriction should be applied not only in the wall-normal direction but also in the streamwise direction. The near-wall resolution requirement clearly limits the application of LES in simulating high Reynolds number flows. Therefore, a modeling strategy is required to solve the practical applications. Instead of modeling every detail in the near-wall region, this study uses a wall function approach to reduce the number of computational cells. Cabot and Moin (2000) derived a set of near-wall damping functions and used them to approximate the eddy viscosity in the first cell adjacent to the wall. The eddy viscosity ν_t is obtained from a mixing-length eddy viscosity model with near-wall damping:

$$\frac{\nu_t}{\nu} = \kappa y_w^+ \left(1 - e^{-y_w^+/A}\right)^2, \qquad (2.32)$$

where $y_w^+ = y_w u_\tau / \nu$ is the distance to the wall in wall units, $\kappa = 0.41$ and $A = 19$ (Cabot and Moin, 2000). The filtered velocity field is assumed to satisfy the no-slip condition on the wall.

3. Numerical Implementation

The filtered continuity and momentum equations are solved by using the finite-volume two-step projection method (Bussmann et al., 2002). The forward time difference method is used to discretize the time derivatives. In order to track the free-surface, the VOF method is used. The VOF method was originally developed by Hirt and Nichols (1981) and improved by Kothe et al. (1999) to second-order accuracy on the free surface reconstruction. The current numerical model is modified from Truchas 1.8.4 developed by Doug Kothe, Jim Sicilian and their Telluride team members at Los Alamos National Laboratory. The original program solves N-S equations by using the projection method with the finite volume discretization. A few new algorithms have been added in order to simulate turbulent free surface flows generated by a moving landslide. The VOF method as well as the moving boundary algorithm will be described in Section 3.2.

In order to simulate the free-surface motion in a fixed grid system, which include both air and water, a volume of fluid function, f, representing the volume fraction of water within a computational cell is introduced. The f value equals to one if the cell is full, zero if empty, and $0 < f < 1$ if the cell is partially filled with water (Fig. 3.1). Since f is conserved and governed by the transport equation, it can be described by:

$$\frac{\partial f}{\partial t} + \frac{\partial \overline{u}_i f}{\partial x_i} = 0. \qquad (3.1)$$

The f value represents the spatial position of the fluid body. It also provides the kinematic information on the fluid.

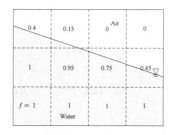

Fig. 3.1. A sketch of the volume fraction f, where $f = 1$ indicates that the cell is filled with water, $f = 1$ filled with air, and $0 < f < 1$ contains an air-water interface.

The momentum equation (2.25) needs to be modified as:

$$\frac{\partial \rho \overline{u}_i}{\partial t} + \frac{\partial \rho \overline{u}_i \overline{u}_j}{\partial x_j} = -\frac{\partial \overline{p}}{\partial x_i} + \rho g_i \\ + \frac{\partial}{\partial x_j}\left[\mu_{eff}\left(\frac{\partial \overline{u}_i}{\partial x_j} + \frac{\partial \overline{u}_j}{\partial x_i}\right)\right], \quad (3.2)$$

where the cell density ρ is defined as:

$$\rho = f \rho_w \quad (3.3)$$

or in the vector form:

$$\frac{\partial \rho \overline{\mathbf{u}}}{\partial t} + \nabla \cdot (\rho \overline{\mathbf{u}}\,\overline{\mathbf{u}}) = -\nabla \overline{p} + \rho \mathbf{g} + \nabla \cdot \left[\mu_{eff}\left(\nabla \overline{\mathbf{u}} + \nabla^T \overline{\mathbf{u}}\right)^n\right], \quad (3.4)$$

where μ_{eff} is the cell viscosity.

3.1. Finite Volume Discretization Method

Consider a general conservation equation for an arbitrary quantity ϕ:

$$\frac{\partial \phi}{\partial t} + \nabla \cdot (\mathbf{u}\phi) = S(\phi), \tag{3.5}$$

where S is a source term which is generally dependent upon ϕ. The finite volume method starts from the volume integral form of the partial differential equation, e.g., (3.5), we have,

$$\int \left[\frac{\partial \phi}{\partial t} + \nabla \cdot (\mathbf{u}\phi) = S(\phi) \right] dV. \tag{3.6}$$

By adopting the Gauss divergence theorem, (3.6) can be expressed as

$$\int \frac{\partial \phi}{\partial t} dV + \oint d\mathbf{A} \cdot (\mathbf{u}\phi) = \int S(\phi) dV, \tag{3.7}$$

where V is the volume, and $\mathbf{A}$ is the area vector

$$\mathbf{A} = \mathbf{n}A, \tag{3.8}$$

with $\mathbf{n}$ being the unit normal vector and A the face area (Fig. 3.2).

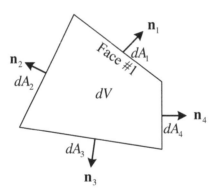

Fig. 3.2. A sketch of the control volume and face normal vectors. Each face has a unique normal vector $\mathbf{n}$ and face area dA.

The physical domain then can be divided into a number of discrete volume elements ("cells") denoted by subscript i. In (3.7), the volume integral can be viewed as the integral over each cell, and the surface integral is over the bounding surface of the cell. Equation (3.7) can then be approximated by a discrete numerical scheme:

$$\frac{\phi_i^{n+1} - \phi_i^n}{dt} + \frac{1}{V_i}\sum_F [\mathbf{A}\cdot\mathbf{u}^n]_F [\phi]_F^n = S_i^n, \qquad (3.9)$$

where superscript n indicates the nth time step, dt is the time step size, V_i is the ith cell volume, and subscript F denotes the cell face.

On the left-hand side of (3.9), the surface integral has been approximated as a sum over discrete cell faces F with an area vector $\mathbf{A}_F$; V_i is the control volume, and ϕ_i is a local volume-averaged value,

$$\phi_i = \frac{\int \phi dV}{\int dV} = \frac{1}{V_i}\int \phi dV, \qquad (3.10)$$

and similarly S_i denotes the volume-averaged source function.

In (3.9), a first-order forward time-differencing scheme has been adopted. For spatially second order schemes, ϕ is assumed to vary linearly in space. Hence, ϕ_i is expressed as:

$$\phi_i = \phi(\mathbf{x}_c), \qquad (3.11)$$

where $\mathbf{x}_c$ is the geometric centroid of the control volume. Similarly, the face quantities $\langle \phi \rangle_F$ are given at the geometric centroid of each face.

Because of the usage of FVM, it is natural to adopt the collocated arrangement that defines all the fluid properties at the cell centroids, including density, pressure, velocities, and viscosity. However, from (3.9), the FVM requires information not only at the cell centroid but also on the face centroid. The face centroid quantities may be a scalar (e.g., density), a vector (e.g., velocity vector), or a gradient (e.g., pressure gradient or velocity gradient). Therefore, an algorithm is required to convert cell centroid data to face centroids or to evaluate the quantity gradients at each cell centroid. The gradient on the cell face is calculated based on the cell-centered values which adjoin the cell face.

3.2. Interface Kinematics

In order to implement the free surface boundary conditions, the position of the free surface must be determined first. The VOF algorithm, developed by Kothe *et al.* (1999), is robust and computational efficient and will be adopted in this study.

The VOF algorithm involves two major steps: the transport of the volume fraction and the interface reconstruction.

3.2.1. *The Transport of the Volume Fraction*

In the VOF algorithm, besides the mass and momentum equations, an additional equation, (3.1), for the volume fraction, f, must be solved. Equation (3.1) is also called the VOF equation and can be discretized, based on FVM formulation:

$$f^{n+1} = f^n - \frac{1}{V_i}\sum_F dt\left[\mathbf{A}\cdot\mathbf{u}^n\right]_F [f]_F^n. \qquad (3.12)$$

The second term in the right-hand side of (3.12) denotes the volume fluxes across the cell faces. However, unless small dt has been used, the error could be large if the volume fluxes are estimated by the volume fraction on cell faces (Fig. 3.3). The volume fluxes can be estimated more precisely based on the geometric calculation:

$$f^{n+1} = f^n - \frac{1}{V_i}\sum_F dt\left[\mathbf{A}\cdot\mathbf{u}^n\right]_F f_F^n, \qquad (3.13)$$

where f_F^n denotes the volume fraction of dV_F^n associated with water through a cell face:

$$f_F^n = \frac{dV_{w,F}^n}{dV_F^n}. \qquad (3.14)$$

Once the solution to (3.13) is found, the fluid volumes are marched forward in time.

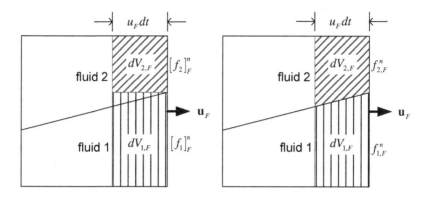

Fig. 3.3. The advected mass through a cell face. The left one tends to have larger error than the right one.

3.2.2. Volume Tracking Algorithm

To solve the volume fraction f^{n+1} at the new time-step from (3.13), we need the information on f_F^n as well as the free surface reconstruction. In this paper, a multidimensional PLIC (piecewise linear interface calculation) (Rider and Kothe, 1998) is utilized to construct the plane in each free surface cell. The plane is then used to estimate f_F^n.

The PLIC algorithm consists of two steps: the planar reconstruction and the calculation of the volume of water across a cell face. As for the first step, the free surface within a cell is assumed to be a plane. Using the information on f^n, the orientation of the free surface can be estimated as the gradient of f^n. As for the second step, the volume of water across a cell face can be evaluated by multiplying the volume flux by dt. This volume is then used to update the volume fraction in the cell and to transport other quantities in the momentum equations. We shall discuss the detailed algorithm in the following sections.

3.2.2.1. Estimation of normal vector on the free surface

A normal vector to the free surface is defined as the gradient of the volume fraction, f, i.e.,

$$\mathbf{n} = \nabla f, \tag{3.15}$$

and can be normalized as:

$$\hat{\mathbf{n}} = \frac{\nabla f}{|\nabla f|}. \qquad (3.16)$$

The cell center and cell face gradient calculations are carried out via the least square algorithms.

By assuming that the free surface geometry is piecewise linear (planar), the free surface in each free surface cell can be described as:

$$\hat{\mathbf{n}} \cdot \mathbf{x} - C_p = 0, \qquad (3.17)$$

where $\mathbf{x}$ is the position vector of a point on the plane and C_p is the plane constant to be determined.

3.2.2.2. Reconstruction of the free surface

After determining the normal vector, $\hat{\mathbf{n}}$, on the free surface, the free surface locations can be calculated iteratively. For each free surface cell, the free surface plane divides space into regions inside and outside of water body, depending upon the direction chosen by $\hat{\mathbf{n}}$. Based on the definition expressed in (3.16), $\hat{\mathbf{n}}$ points to the interior of the water body. Hence, $\hat{\mathbf{n}} \cdot \mathbf{x} - C_p$ will be positive for any point inside the water body; zero for any point $\mathbf{x}$ on the free surface plane; and negative for any point $\mathbf{x}$ outside the water body. The volume V_{tr} is the truncation volume lying within the water body and separated by (3.17). By varying C_p, V_{tr} will also be changed. Our goal is to find the constant C_p iteratively (see Fig. 3.4) so that a nonlinear function defined as:

$$E(C_p) = V_{tr}(C_p) - f * V \qquad (3.18)$$

becomes zero or smaller than a convergence criteria. The algorithm of geometric calculation of the truncation volume V_{tr} can be found in Rider *et al.* (1998) and Kothe *et al.* (1999).

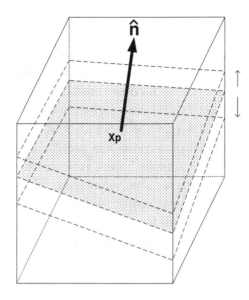

Fig. 3.4. Locating the interface. The interface can be moved up and down by changing C (Eq. (3.17)) and followed the direction of $\hat{n}$. This is constrained by the volume conservation.

When $E(C_p)$ is equal to zero or smaller than a certain tolerance, the free surface plane is declared "reconstructed" in that cell. There are many root-finding algorithms, e.g., Bisection, Newton's method, and Brent's method, are available to find the zero of this nonlinear function, but Rider and Kothe (1998) have found that Brent's method (Press *et al.*, 1986) gives the best results in practice.

After the free surface plane is reconstructed, a geometric calculation of volume fluxes of water across cell faces can be done based on the plane equation (3.17) of each cell.

Now our volume tracking algorithm template is summarized as follows:

1. Estimate the interface normal $\hat{n}$ from discrete f data.
2. Find the plane constant C_p in (3.17) by solving (3.18) to reconstruct the interfaces within the cell in a volume conservation manner.
3. A geometric calculation of the volume fluxes of different materials crosses cell faces based on f_F^n.
4. Update the cell center volume fraction from (3.13).

3.2.3. *Treating Air as Void*

In our simulations, because both the pressure gradient and the momentum flux are often much smaller in the air region than those in the water region, a void model is used to simplify the equations in the air region. Voids are used to represent regions where the cells are fully occupied by air. If a cell is void, the N-S equations are not solved and the pressure in the void regions is a constant or zero.

Once the void model is applied, there is no momentum contribution from air. Therefore, the there is no momentum exchange between the void and non-void cells. Consequently, the face flux, e.g., the momentum flux through the cell face, will be set to zero, if this cell face is adjacent to a void cell. If a non-void cell is connected to a void cell, the velocity gradient on the cell face is then calculated from the neighboring cells with fluid in them.

3.3. Projection Method

The projection method has been widely used to solve the NS equations. This method was first proposed by Chorin (1968) and then used by Bell and his coworkers to solve the constant-density (Issa, 1986) or variable-density (Rider and Kothe, 1998) incompressible N-S equations. In this study, the projection method will be used to solve the filtered N-S equations.

In the projection method, momentum equations, (3.2), can be described by two fractional steps:

$$\frac{\rho^{n+1}\overline{\mathbf{u}}^* - \rho^n\overline{\mathbf{u}}^n}{dt} = -\nabla\cdot\left(\rho^n\overline{\mathbf{u}}\,\overline{\mathbf{u}}\right)^n + \nabla\cdot\left(\mu_{eff}^n\left(\nabla\overline{\mathbf{u}} + \nabla^T\overline{\mathbf{u}}\right)^n\right), \quad (3.19)$$

$$\frac{\rho^{n+1}\overline{\mathbf{u}}^{n+1} - \rho^{n+1}\overline{\mathbf{u}}^*}{dt} = -\nabla\overline{p}^{n+1} + \rho^{n+1}\mathbf{g}, \quad (3.20)$$

where (3.19) is an explicit expression for the interim velocity $\overline{\mathbf{u}}^*$, referred to as the predictor step. In (3.19), all forces except gravity and pressure gradient are included. On the other hand, (3.20) is called the projection step. Combining (3.19) with (3.20) produces the time discretization of (3.2):

$$\frac{\rho^{n+1}\bar{\mathbf{u}}^{n+1} - \rho^n\bar{\mathbf{u}}^n}{dt} = -\nabla\cdot(\rho\bar{\mathbf{u}}\bar{\mathbf{u}})^n - \nabla\bar{p}^{n+1} + \rho^{n+1}\mathbf{g} \\ + \nabla\cdot\left(\mu_{eff}^n\left(\nabla\bar{\mathbf{u}}^n + \nabla^T\bar{\mathbf{u}}\right)^n\right). \quad (3.21)$$

No additional approximation results from this decomposition.

Equation (3.20) relates $\bar{\mathbf{u}}^{n+1}$ to $\bar{\mathbf{u}}^*$. By adopting the solenoidal condition, (2.18), we have:

$$\nabla\cdot\frac{\nabla\bar{p}^{n+1}}{\rho^{n+1}} = \nabla\cdot\left(\frac{\bar{\mathbf{u}}^*}{dt} + \mathbf{g}\right). \quad (3.22)$$

The above equation is also called the Poisson Pressure Equation (PPE). The cell-centered pressure $\bar{p}^{n+1}$ at the new time step can be obtained by solving (3.22), and cell-face pressure gradient $\nabla\bar{p}_F^{n+1}$ is calculated from the same stencil as that of the density interpolation to faces. Equation (3.20) is then used to calculate the solenoidal face velocity field:

$$\bar{\mathbf{u}}_F^{n+1} = \bar{\mathbf{u}}_F^* - dt\left(\frac{\nabla\bar{p}_F^{n+1}}{\rho_F^{n+1}} - \mathbf{g}\right). \quad (3.23)$$

Finally, the pressure gradient on cell faces $\nabla\bar{p}_F^{n+1}/\rho_F^{n+1}$ is interpolated to cell centers in order to obtain the velocity field $\bar{\mathbf{u}}^{n+1}$ at new time-step, which satisfies the divergence free condition.

3.3.1. *Momentum Advection*

After obtaining the volume tracking results, the momentum advection $\nabla\cdot(\rho\bar{\mathbf{u}}\bar{\mathbf{u}})^n$ can be evaluated by the face-centered velocities. Discretizing the advection term of (3.19), we obtain:

$$dt \int \nabla \cdot \langle \rho \bar{\mathbf{u}} \bar{\mathbf{u}} \rangle^n \, dV \approx \sum_F dV_F^n \langle \rho \bar{\mathbf{u}} \rangle_F^n$$
$$= \sum_F M_F^n \bar{\mathbf{u}}_F^n, \quad (3.24)$$

where dV_f is the advected volume through each cell face, and M_F^n is the advected mass through each cell face. Figure 3.1 illustrates the advection of water and air across the right face of a cell containing the free surface. The advected volume is:

$$dV_F = dt \mathbf{A}_F \cdot \bar{\mathbf{u}}_F = dt A_F \bar{\mathbf{u}}_F \cdot \overline{\hat{\mathbf{n}}}_F, \quad (3.25)$$

where $\bar{\mathbf{u}}_F \cdot \overline{\hat{\mathbf{n}}}_F$ is the normal component of the solenoidal face velocity $\bar{\mathbf{u}}_F$ on the face, and A_F is the face area.

The mass of water leaving a cell across face F can be expressed by multiplying $dV_{w,F}$ and the corresponding densities:

$$M_F = \rho_w dV_{w,F}, \quad (3.26)$$

and with (3.14):

$$M_F = \rho_w dV_F f_F. \quad (3.27)$$

Equation (3.24) can then be written as:

$$dt \int \nabla \cdot (\rho \mathbf{u} \mathbf{u})^n \, dV \approx \sum_F M_F^n \bar{\mathbf{u}}_F^n$$
$$= dt \sum_F \rho_w \left(\mathbf{A}_F \cdot \bar{\mathbf{u}}_F^n \right) f_F^n \bar{\mathbf{u}}_F^n, \quad (3.28)$$

where $\left(\mathbf{A}_F \cdot \bar{\mathbf{u}}_F^n \right) f_F^n$ is directly obtained from (3.13). Thus, (3.24) is consistent with mass advection. Note that (3.28) is an upwind scheme and is a first-order accurate approximation.

3.3.2. *Momentum Diffusion*

The net viscous stress on the control volume is calculated by applying the divergence theorem to the volume integral of the local stress. This reduces to a sum of the dot product of the face normal vector with the local velocity gradient multiplied by the face area:

$$\nabla \cdot \left(\mu_{eff}^n \left(\nabla \bar{\mathbf{u}} + \nabla^T \bar{\mathbf{u}} \right)^n \right) \\ = \sum_F \mu_{eff,F}^n A_F \left[\bar{\hat{\mathbf{n}}}_F \cdot \left(\nabla \bar{\mathbf{u}}_F + \nabla^T \bar{\mathbf{u}}_F \right) \right]^n. \quad (3.29)$$

The velocity gradient is calculated by the least squares method. The effective cell center viscosity μ_{eff} is evaluated by using volume weighted technique.

$$\mu_{eff} = f \mu_{w,eff}. \quad (3.30)$$

The cell center viscosity is then harmonically averaged to face centroids by using an orthogonal approximation:

$$\mu_{eff,F} = \frac{2}{1/\mu_{nb} + 1/\mu_p}, \quad (3.31)$$

where μ_{nb} is the neighbor cell viscosity and μ_p is the cell center viscosity.

3.3.3. Projection

After obtaining the momentum advection and diffusion, the "predicted" velocity $\bar{\mathbf{u}}^*$ can be calculated from (3.19). The predicted velocity $\bar{\mathbf{u}}^*$ and f^{n+1} are then interpolated to cell faces to obtain $\mathbf{u}_F^*$ and ρ_F^{n+1}, respectively. Since we assume velocity does not vary discontinuously near the free surface, the velocity interpolation can be done by the least square linear reconstruction method. However, because density varies discontinuously across the free surface, we have to limit the size of the stencil. The cell face densities are calculated by averaging the cell densities on either side of the face.

The discretization of (3.22) is:

$$\sum_F \left(\mathbf{A}_F \cdot \frac{\nabla \bar{p}_F^{n+1}}{\rho_F^{n+1}} \right) = \sum_F \left(\mathbf{A}_F \cdot \left(\frac{\bar{\mathbf{u}}_F^*}{dt} + \mathbf{g}_F \right) \right), \qquad (3.32)$$

where the cell face pressure gradient $\nabla \bar{p}_F^{n+1}$ is calculated from the same stencil as that of the density interpolation to faces:

$$\nabla \bar{p}_F^{n+1} = \frac{\bar{p}_{nb}^{n+1} - \bar{p}_p^{n+1}}{dx}. \qquad (3.33)$$

Here $\bar{p}_{nb}^{n+1}$ is the neighbor cell pressure, $\bar{p}_p^{n+1}$ the cell pressure, and δx the centroid-to-centroid distance.

Equation (3.32) is first solved for the cell-centered pressure $\bar{p}^{n+1}$. Equation (3.23) is then used to calculate the solenoidal face velocity field $\bar{\mathbf{u}}_F^{n+1}$. The last step is to interpolate $\nabla \bar{p}_F^{n+1} / (\rho_F^{n+1}) - \mathbf{g}$ to cell centers, which, along with (3.20) and $\bar{\mathbf{u}}^*$, is used to obtain cell center velocity field $\bar{\mathbf{u}}^{n+1}$.

Equation (3.32), the PPE, is a set of linear algebraic equations for the pressure field that can be solved by iterative methods. In this study, the conjugate gradient (CG) algorithm with symmetric successive over-relation (SSOR) (Varga, 1962) pre-conditioner (Golub and Van Loan, 1989) will be used to solve the linear algebraic equations.

3.4. Computational Cycle

We now summarize the computing cycle to update the field variables within one time step. Given values of $\bar{\mathbf{u}}^n$, $\bar{\mathbf{u}}_F^n$, and $\bar{p}^n$ at the time step n, we advance the solutions to the time step $n+1$ in the following procedures:

1. Solve (3.13) for f^{n+1} using $\bar{\mathbf{u}}_F^n$ and utilizing the multidimensional PLIC volume tracking algorithm. New time step cell density ρ^{n+1} is obtained via $\rho^{n+1} = f^{n+1} \rho_w$ and a geometric calculation of volume fluxes of water crosses cell faces.
2. Obtain $\bar{\mathbf{u}}^*$ by (3.19) where $\nabla \cdot (\rho \bar{\mathbf{u}} \bar{\mathbf{u}})^n$ is evaluated by using the volume flux information obtained directly from the previous step, and the viscous force is obtained via (3.29).

3. $\bar{\mathbf{u}}_F^*$ and ρ_F^{n+1} are obtained by interpolating $\bar{\mathbf{u}}^*$ and ρ^{n+1} to cell faces.
4. Solve PPE, (3.22) and (3.32), to obtain cell center pressure p^{n+1}.
5. A solenoidal face velocity field $\bar{\mathbf{u}}_F^{n+1}$ is obtained by (3.23).
6. Finally, ($\nabla \bar{p}_F^{n+1} / \rho_F^{n+1} - \mathbf{g}_F$) is interpolated to cell center to obtain $\bar{\mathbf{u}}^{n+1}$ from $\bar{\mathbf{u}}^*$ via (3.20).

3.5. Moving Solid Algorithm

In order to simulate the movement of a solid body through fixed mesh, the partial grid algorithm and an internal source function are added to the program. Consider a volume V containing an obstacle (Fig. 3.5), which is defined by its volume $V_{obst(t)}$ and its surface $A_{obst(t)}$. If the obstacle volume increases, i.e. $dV_{obst}/dt > 0$, the volume of the fluids decreases, and vice versa. The conservation of mass in the volume V can be expressed as below:

$$\oint_A \mathbf{u} \cdot \mathbf{n} dA = \frac{dV_{obst(t)}}{dt}. \qquad (3.34)$$

The new continuity equation for the volume V can be expressed as:

$$\nabla \cdot \bar{\mathbf{u}} = \frac{1}{V}\frac{dV_{obst(t)}}{dt} = \phi(x, y, z, t), \qquad (3.35)$$

where ϕ = internal source function. Substituting the new continuity equation, (3.35), into the original governing equations, the momentum equation becomes:

$$\frac{\partial(\rho \bar{u}_i)}{\partial t} + \frac{\partial(\rho \bar{u}_i \bar{u}_j)}{\partial x_j} = -\frac{\partial \bar{p}}{\partial x_i} + \rho g_i \\ + \frac{\partial}{\partial x_j}\left[\mu_{eff}\left(\frac{\partial \bar{u}_i}{\partial x_j} + \frac{\partial \bar{u}_j}{\partial x_i}\right)\right] + \overline{\rho u_i}\phi. \qquad (3.36)$$

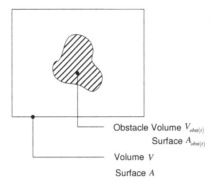

Fig. 3.5. A solid body in the fluid domain.

The main procedure of the moving solid algorithm can be expressed as:

1. Update the new time step solid body VOF (Fig. 3.6).
2. Add a positive source function to the region where water will be pushed out, and a sink term to the wake zone where water will be sucked in (Fig. 3.7).
3. Solve the new time step velocities.

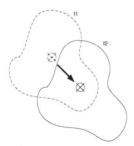

Fig. 3.6. Update the solid body VOF from old time step (t1) to new time step (t2).

The difference between present moving solid algorithm and Heinrich's (1992) is that the present algorithm updates the new time step solid VOF first, and then adds additional source functions to the areas where solid materials change. In Heinrich's method, the source functions are added before updating the solid VOF. The present algorithm yields more accurate moving boundary effects (Fig. 3.8).

One of the advantages of using the moving solid algorithm is that the grid system does not need to fit the solid boundary. Therefore, a part of the internal cells will be occupied by the solid material. In order to deal with a cell occupied partially by solid materials and water, a simple partial cell treatment will be applied.

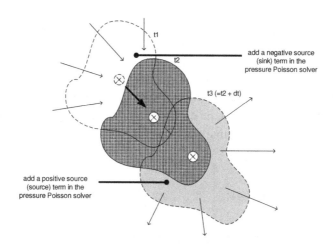

Fig. 3.7. Add a positive source term in front of the solid body, and a negative source term (sink term) in the wake zone. Where dt = t2 − t1, t3 = t2 + dt.

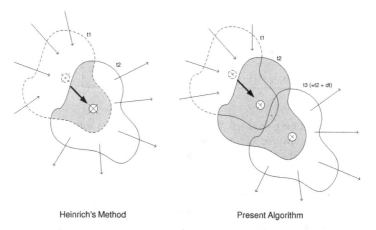

Fig. 3.8. The comparison between Heinrich's method and present moving solid algorithm.

3.5.1. Partial Cell Treatment

The purpose of the partial cell treatment is to adopt the moving solid algorithm and to deal with unstructured interior obstacles. Since the obstacle inside the computational domain occupies a part of the interior fluid volume, the effective cell volume will be a fraction of the original cell volume:

$$V_{eff} = (1 - f_{solid})V \\ = \theta V, \tag{3.37}$$

where V_{eff} is the effective cell volume, f_{solid} is the volume fraction occupied by solid material in each cell, V is the original cell volume, and θ is the effective volume fraction (Fig. 3.9).

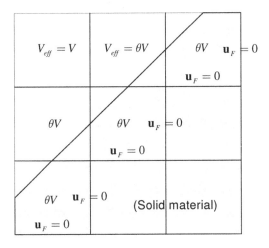

Fig. 3.9. A sketch of the partial cell treatment. If a cell contains partial solid material, the effective cell volume will be a fraction of the original cell volume. The face velocity will be set to zero if at least one side of the cell is fully filled with solid material.

If a cell contains partially water and solid material, the flow solver has to deal with it. Cell faces are defined either to be entirely closed or opened. Cell faces are "closed" only if at least one of the two immediately neighboring cells is entirely occupied by solid material. If the cell faces are "closed", the face velocity of the cell is set to zero, and the face pressure is no longer calculated in the pressure solution. On the other hand, if any face between two cells, containing at least a partial cell volume of fluid, is "open", the code solves the velocities and pressure gradients.

Compared to the conventional way to treat the irregular solid material that creates the "saw-tooth" boundary, the partial cell treatment introduced here is a better choice to represent the smooth real geometry of the boundary.

4. Landslide Simulations and Benchmark Problems

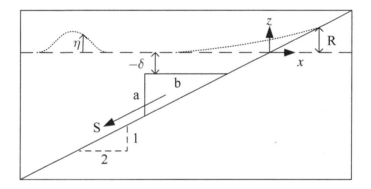

Fig. 4.1. A definition sketch for the physical variables employed in this study.

4.1. *Channel Description*

A series of landslide experiments was carried out in a wave tank with a length of 104 m, width of 3.7 m, and depth of 4.6 m at Oregon State University (Synolakis and Raichlen, 2003; Raichlen and Synolakis, 2003). A 1:2 (vertical to horizontal) uniform slope was installed near one end of the tank. A solid wedge was used to represent the landslide and to generate waves (Fig. 4.1). The triangular face has the following dimensions: a length of b = 91.44 cm, a front face a = 45.72 cm and a width of w = 61 cm. The wedge moved down the slope by gravity. The experiments provided several pieces of information, such as the initial wedge elevation δ, specific weight (defined as $\gamma = \rho_{wedge}/\rho_w$), the trajectory of the wedge movements, time-history of runup, and time-history of free-surface elevations. Figure 4.2 shows the positions of runup gauges and wave gauges. One of the submerged landslide cases has been used as the benchmark problem #4 discussed in the Third International Workshop on Long-Wave Runup Models held at Wrigley Marine Science Center, Catalina Island, California (2004).

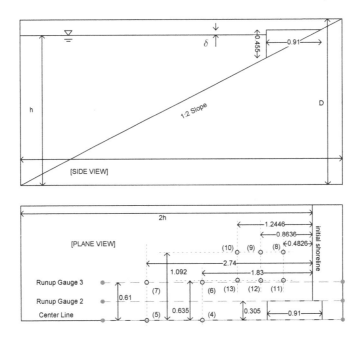

Fig. 4.2. The numerical setup and gauge positions. Unit is in meter.

4.2. *Numerical Simulations*

The numerical simulations are carried out in a numerical wave tank and the landslide is modeled by the moving solid algorithm. Because this study focuses on the runup and water waves generated in the near field, the length of the computational domain is shorter than the laboratory setup (6.6 m). Furthermore, since the vortex shading behind the moving wedge is small, we assume that the flow field is symmetric with respect to the centerline and thus, only half of the physical domain is simulated (Fig. 4.2). The number of grid point is 60 in the streamwise (the direction of landslide movement) direction, 50 in the span-wise direction, and 60 in the vertical direction. A non-uniform grid is used in the streamwise and vertical directions. The finest cell is located at the onshore corner to have a finer resolution for the runup simulations. Coarser resolution is applied to the offshore deep water corner. The side walls as well as the end-wall are treated as impermeable free-slip walls. Therefore, the landslide generated waves will be reflected by the end-wall, which does not exist in the laboratory experiments. However, since the maximum and minimum runup will occur before the reflected waves reach the domain of interest, the reflecting end-wall boundary condition is still a good choice to simplify the numerical setup. Other types of boundary conditions such as numerical sponge layer and advective open boundary condition (Liu *et al.*, 2005) can also be used to avoid the reflection from the end-wall. The boundary condition on the ceiling of the computational domain is the zero-pressure open boundary condition. The 1:2 slope is designed to extend from upper-right corner to lower-left corner (Fig. 4.2), so the solid material will occupy 50% of the cell volume on the slope. This design ensures that the effective cell volume will not be smaller than 50% of the total cell volume, therefore the time marching will not be limited by certain cells with extremely small effective cell volumes due to the Courant number restriction. This design also helps us to clearly identify the shoreline location without being interfered with the irregular effective cell volume. The time step is determined dynamically based on the Courant number criteria, which is 0.45 in this study.

Similar to the laboratory setup, three runup gauges are installed close to the moving landslide; four wave gauges numbered from #4 to #7 are installed in the offshore direction; and one wave gauge array with six wave gauges numbered from #8 to #11 is installed on the lateral side of the moving wedge (Fig. 4.2). However, the gauge locations might not coincide with grid centroid. Thus, the nearest neighbor interpolation method is used to interpolate gauge data. Finally, the displacement of the moving landslide is obtained from the laboratory measurements. Figure 4.3 shows the time history of the displacement (S) and speed of the slide motion for the subaerial landslide case.

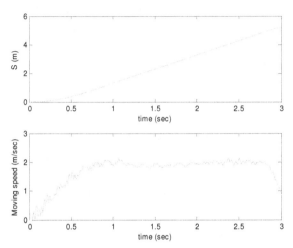

Fig. 4.3. The displacement (S) and the speed of the moving slide. In this subaerial case, the initial wedge position (δ) is 0.454 m above the still water surface. The specific weight (γ) of the wedge is 2.64.

In this subaerial landslide case, the initial elevation of the wedge is 0.454 m above the initial still water level ($\delta = 0.454\text{m}$). The specific weight is $\gamma = \rho_{soild}/\rho_{water} = 2.64$. The numerical simulations are conducted in a domain with size $(x, y, z)_{\text{Domain size}} = (6.6, 1.85, 3.3)$, where the unit is meter and (x, y, z) represent the directions of off-shore (x), lateral (y), and vertical (z) respectively. Non-uniform grids are used in x and z directions with the finest grid size $(dx, dz)_{finest} = (0.0391, 0.0196)$, located at the corner near the initial shoreline. A uniform grid was adopted in the lateral direction. The grid has been specially designed so that $dx = 2dz$. It took about 10 hours (wall clock) to finish the simulation from time = 0.0 sec to 4.0 sec on a personal computer with an Intel 2.8 GHz CPU.

4.3. Results and Discussions

4.3.1. *The Runup Gauges*

In the laboratory setup, two runup gauges were installed on the slope to provide the time history of the runup information. The first runup gauge (Gauge #2) is located at the edge of the sliding wedge, the other one is located at one wedge width away from the centerline. Figure 4.4 shows the comparison between the numerical results and laboratory data. The solid lines are the numerical results and the broken lines are the experimental measurements. The numerical runup heights are determined by the contour line where the water occupies 50% of the effective cell volume. A very good agreement is shown at Gauge #2. However, at Gauge #3 the numerical solutions slightly over-predict the maximum runup heights and the disagreement is about 10%. Overall, the numerical simulation is able to capture the maximum runup heights in the near field region, and most importantly the model is able to accurately predict the arrival time of the maximum runup height.

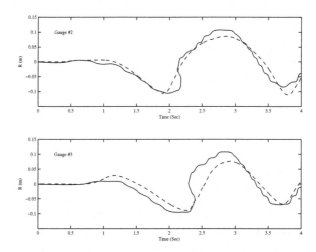

Fig. 4.4. The comparison between numerical results (solid lines) and experimental data (broken lines) for the time history runup height at Gauge #2 and Gauge #3. $\delta = 0.454$ m. $\gamma = 2.64$.

4.3.2. *Four Fixed Wave Gauges*

There are four fixed wave gauges installed in front of the landslide and they are labeled as gauge #4 ~ #7 (see Fig. 4.2). Figure 4.5 shows the comparison of numerical results and wave gauge data. Again, the solid lines are the numerical solutions and the broken lines are the measurements. Gauge #4 and #6 are the ones closer to the shoreline. The comparison shows that the numerical simulation successfully predicts the leading wave height as well as the phase speed. Gauge #5 and #7 show that the numerical solutions under-predict the leading wave heights. However, the numerical model is able to predict the phase of the propagating waves.

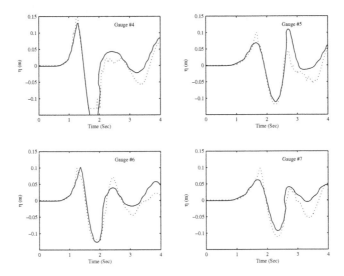

Fig. 4.5. The comparison between numerical results (solid lines) and experimental data (broken lines) for the time histories of free surface fluctuations at wave gauge #4 ~ #7; $\delta = 0.454$ m, $\gamma = 2.64$. The coordinates for gauges are: Gauge #4: $(x, y) = (1.83, 0)$; Gauge #5: $(x, y) = (2.74, 0)$; Gauge #6: $(x, y) = (1.83, 0.61)$; Gauge #7: $(x, y) = (2.74, 0.61)$. The unit is in meter.

4.3.3. Wave Gauge Array

Wave gauges #8 ~ #13 are installed on the side of the moving slide to record the lateral movement of landslide-generated waves. Figure 4.6 shows both the numerical results and laboratory data. The comparisons show that the numerical solutions successfully predict the wave heights of the leading and secondary waves. The numerical solutions also accurately capture the phase of the waves. However, from gauge #8, #9, and #11, around time = 1.75 sec, the comparisons show that the numerical solutions cannot predict some small scale motions. This might be cause by the lack of fine resolution in numerical simulations. It could also be the experimental errors because there are significant wave breaking and air bubble entrainment. Fortunately, these errors do not seem to affect the predictions of the secondary waves.

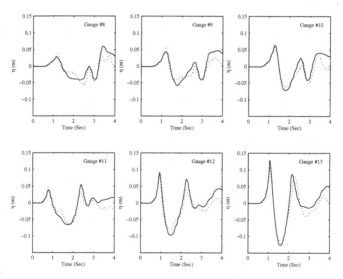

Fig. 4.6. The comparison between numerical results (solid lines) and experimental data (broken lines) for the time histories of free surface fluctuations at wave gauge #8 ~ #13; $\delta = 0.454$ m, $\gamma = 2.64$. The coordinates for gauges are: Gauge #8: (x, y) = (0.4826, 1.092); Gauge #9: (x, y) = (0.8636, 1.092); Gauge #10: (x, y) = (1.2446, 1.092); Gauge #11: (x, y) = (0.635, 0.4826); Gauge #12: (x, y) = (0.635, 0.8636); Gauge #13: (x, y) = (0.635, 1.2446). The unit is in meter.

4.3.4. Free-Surface Profiles

Figure 4.7 shows the snapshots of the free surface profiles. The initial shoreline is at the intersection of $x=0$ and $z=0$. The landslide first pushes the water in front and generates the outgoing waves (time = 0.0 ~ 0.8 sec). Then the landslide submerges into water and the free surface caves in, generating strong lateral converging flows (time = 0.8 ~ 1.2 sec). While the landslide keeps submerging, the converging lateral flows in the wake zone generate strong reflecting (rooster-tail) waves (time = 1.2 ~ 3.0 sec). These rooster-tail waves are the key source to the maximum runup. After time = 3.0 sec, the solution is gradually contaminated by the waves reflected from the side walls and will be excluded from this discussion.

4.3.5. Velocities

Figure 4.8 presents the detailed velocity vectors at the centerline cross-section. From time = 0.0 ~ 0.8 sec, the velocities are generated by the entry of the landslide. After time = 0.8 sec, the elevation of the top of the slide is lower than the still water surface and water starts to flood into the wake area above the slide. At time = 1.2 sec, the slide is fully submerged. A very complex three-dimensional flow pattern can be observed right after the slide has fully submerged into water (time = 1.2 ~ 1.5 sec). The depth-integrated equation models are not suitable for modeling this type of flows. From time = 2.1 to 2.7 sec, the velocity distributions show that a strong flow current has been generated by the slide motion, which has a thickness about 1.5 ~ 2.0 times the front face height (H_{slide}) of the slide. Above this current, there exists a returning current and generates a negative wave. Between these two currents, there exists an eddy that can be clearly seen from time = 2.1 sec to 2.7 sec. The size of the eddy depends on the moving speed of the slide. In this case, the eddy size is about one front face height (H_{slide}) of the slide. After t = 2.7 sec, the slide stops. However, the current behind the slide still keeps on moving with the decreasing strength.

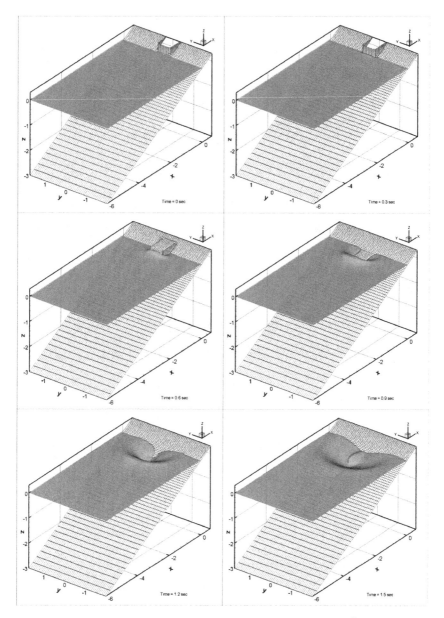

Fig. 4.7. Snapshots of free-surface profile for the sliding wedge with $\delta = 0.454$ m, and $\gamma = 2.64$. The unit is in meter.

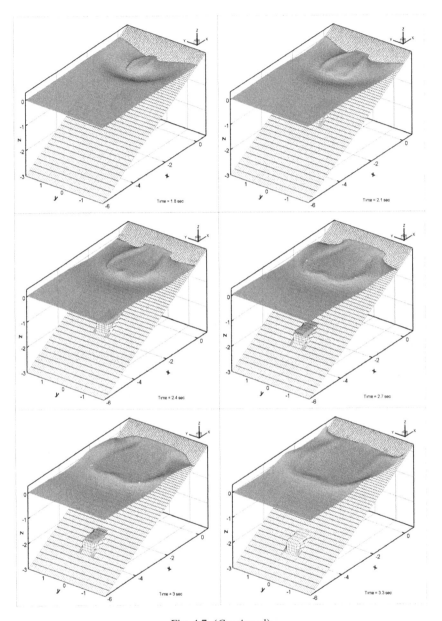

Fig. 4.7. (*Continued*)

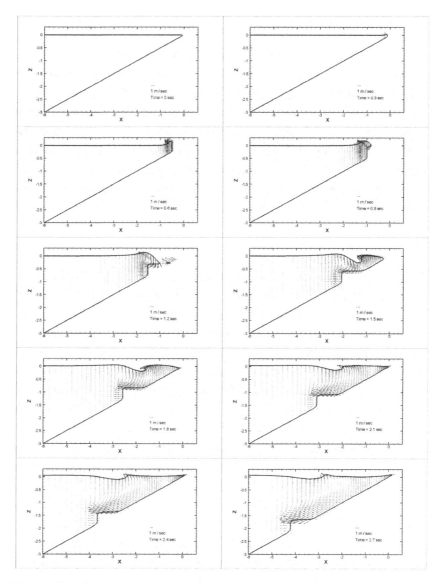

Fig. 4.8. Snapshots of velocity vectors on the centerline vertical plane for the sliding wedge with $\delta = 0.454$ m, and $\gamma = 2.64$. The unit is in meter.

4.3.6. Shoreline Movement

Figure 4.9 shows the snapshots of the shoreline movement. The snapshot at time = 1.2 sec shows that the slide has fully submerged into water and an air bubble is trapped there. After time = 1.5 sec, the shoreline starts to move up, spreads out, and then reaches the maximum runup height at time = 2.7 sec. From the sequence of the snapshots we observe that at time = 2.7 sec, both the centerline and the near field region (-1 m $\leq y \leq 1$ m) reaches the maximum runup. Therefore, the intersection of the shoreline and $z = 0$ m at time = 2.7 sec can be treated as the maximum inundation area.

4.3.7. x-Cross-Section

Figure 4.10 provides the velocity distribution from $x = -2.5$ m to $x = -0.25$ m, where the initial shoreline is located at $(x, z) = (0, 0)$. On each plot, the vectors are the fluid particle velocity vectors, and the dot lines are the water boundaries. The upper solid lines indicate the interface between air and water, and the lower solid lines indicate the interface between water and solid walls. Starting from $x = -2.5$ m, the water is pushed away and moves into the ambient water body. The region affected by the moving slide has a length scale about 3 times the width of the slide (W_{slide}) laterally and 2 times the height of the slide (H_{slide}). From $x = -2.0$ m to $x = -1.5$ m, the cross-sections are interacting with the slide. An important feature is that a strong negative wave has been generated by the slide. At these cross-sections, water is generally moving away from the slide. However, starting from $x = -1.75$ m, the converging currents at both sides of the slide can be observed. These converging currents become significantly and dominate the flow domain from $x = -1.25$ m to $x = -0.25$ m. These converging currents collide into each other and generate a strong rooster-tail wave.

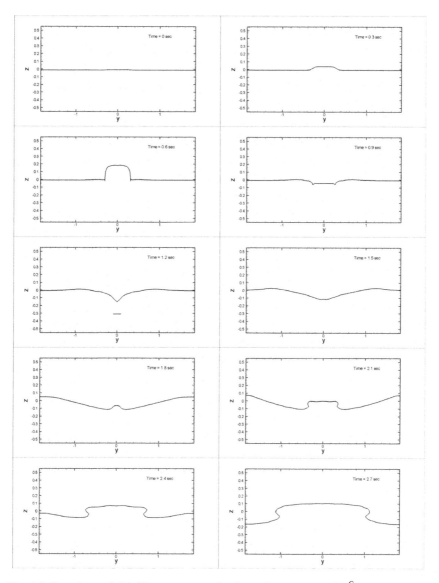

Fig. 4.9. Snapshots of shoreline movement for the sliding wedge with $\delta = 0.454$ m, and $\gamma = 2.64$. The unit is in meter.

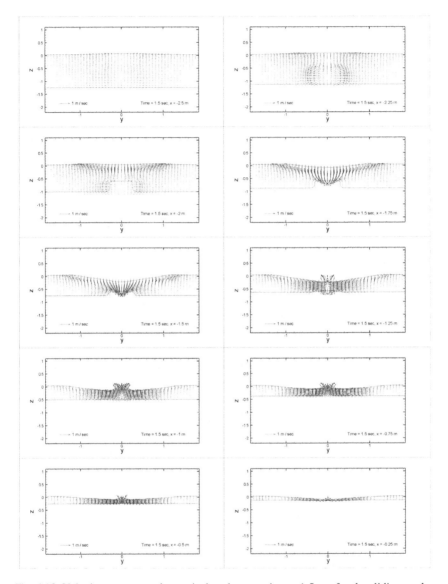

Fig. 4.10. Velocity vectors on the vertical x-planes at time = 1.5 sec for the sliding wedge with $\delta = 0.454$ m, and $\gamma = 2.64$. The unit is in meter. The magnitude of the reference vector indicates the speed of the wedge.

4.3.8. y-Cross-Section

Figure 4.11 shows the cross-section from $y = 0.0$ m to $y = 0.8$ m, where $y = 0.0$ m denotes the centerline cross-section. From $y = 0.0$ m to $y = 0.3$ m, the velocity distributions clearly show how the water has been pushed away from the moving slide, and how the water has been dragged by the slide in the wake zone. At $y = 0.6$ m and $y = 0.8$ m, the converging currents can be also observed above the slope.

4.3.9. z-Cross-Section

Figure 4.12 shows the cross-section from $z = -1.1$ m to $z = 0.0$ m. Starting from $z = -1.1$ m, the plot shows the region affected by the slide movement. Clearly, the largest velocity happens in front of the slide and within one slide-width, W_{slide}. From $z = -0.8$ m to $z = -0.2$ m, the details of converging flows in the wake area are evident. At $z = -0.1$ m and $z = 0.0$ m, the complex flow patterns can be seen and the fluid motion is mainly toward the offshore direction.

5. A Submerged Landslide Simulation

In the previous section, we have verified the accuracy of a 3D subaerial landslide simulation. In this section, a 3D submerged landslide simulation is going to be inspected and discussed. Because there are many similarities between the subaerial and submerged landslide cases, we shall focus the descriptions and discussions on the differences between these two cases. We note that this is also the benchmark problem #4 discussed in the workshop.

5.1. Setup

The initial elevation of the submerged landslide is 0.05 m below the still water level ($\delta = -0.1$ m). The specific weight γ is 2.64. The numerical simulation is conducted in a domain with size $(x, y, z)_{\text{Domain size}} = (5.1, 1.85, 2.55)$. In order to save computational

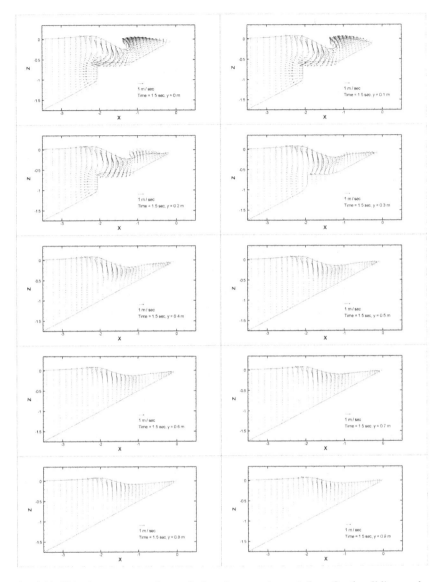

Fig. 4.11. Velocity vectors on the vertical y-planes at time = 1.5 sec for the sliding wedge with $\delta = 0.454$ m, and $\gamma = 2.64$. The unit is in meter. The magnitude of the reference vector indicates the speed of the wedge.

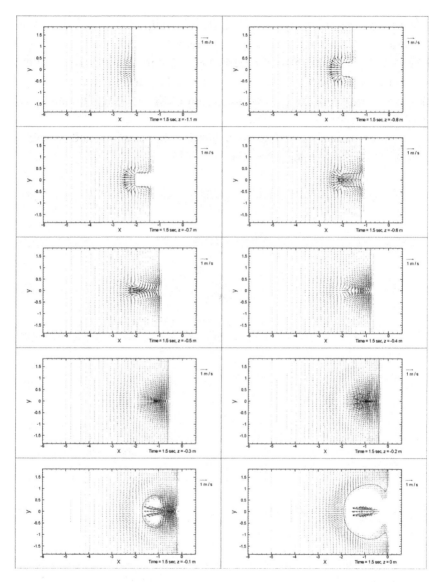

Fig. 4.12. Velocity vectors on the horizontal z-planes at time = 1.5 sec for the sliding wedge with δ = 0.454 m, and γ = 3.24. The unit is in meter. The magnitude of the reference vector indicates the speed of the wedge.

efforts, we apply the symmetric boundary condition at $y = 0.0\,\text{m}$. Non-uniform grids are applied in x and z directions with the finest grid size $(dx, dz)_{finest} = (0.02050,\ 0.01025)$ located at the corner near the initial shoreline. A uniform grid is adopted in the lateral direction. Figure 5.1 shows the displacement (S) and the speed of the moving slide measured from the laboratory experiment.

The submerged landslide case has the same initial and boundary conditions as those in the subaerial case.

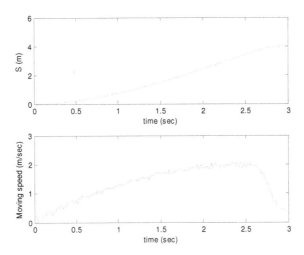

Fig. 5.1. The displacement (S) and the speed of the moving slide. In this subaerial case, the initial wedge position (δ) is −0.1 m above the still water surface. The specific weight (γ) of the wedge is 2.64.

5.2. Results and Discussions

5.2.1. *The Runup Gauges*

Figure 5.2 shows the comparisons of the numerical solutions and laboratory data. From the comparison of Gauge #2 and #3, the numerical results appear to be in good agreement with the laboratory data in terms of the runup height and phase. Again, this comparison validates the accuracy of the numerical model.

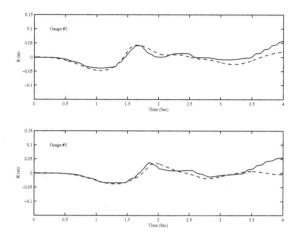

Fig. 5.2. The comparison between numerical results (solid lines) and experimental data (broken lines) for the time history runup height at Gauge #2 and Gauge #3. $\delta = -0.1$ m. $\gamma = 2.64$.

5.2.2. *Four Fixed Wave Gauges*

Figure 5.3 shows the comparisons of wave gauge data. Gauge #4 ~ Gauge #7 are placed in the offshore region. The comparisons show a very good agreement between the numerical solutions and the laboratory data. The numerical model successfully predicts the maximum wave height in Gauge #6 and Gauge #7, and captures the phases of waves in all measurements.

From Gauge #4 to Gauge #7, the maximum wave height does not happen at the leading but at the secondary wave. This observation is different to that in the subaerial case. This shows that in the subaerial landslide, a large portion of energy is transferred from slide to water by the initial "pushing" process. However, in the submerged landslide case, the momentum generated by the initial pushing process will be transferred to the wake area, and has fewer effects on the free-surface waves.

5.2.3. *Wave Gauge Array*

Figure 5.4 shows the comparison of Gauge #8 ~ Gauge #13. These six wave gauges mainly monitor the energy transfer in the lateral direction.

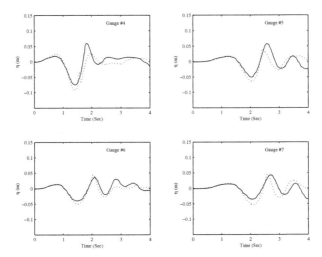

Fig. 5.3. The comparison between numerical results (solid lines) and experimental data (broken lines) for the time histories of free surface fluctuations at wave gauge #4 ~ #7; $\delta = -0.1$ m, $\gamma = 2.64$. The coordinates for gauges are: Gauge #4: (x, y) = (1.83, 0); Gauge #5: (x, y) = (2.74, 0); Gauge #6: (x, y) = (1.83, 0.61); Gauge #7: (x, y) = (2.74, 0.61). The unit is in meter.

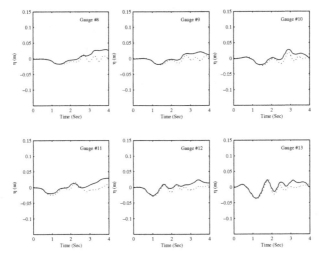

Fig. 5.4. The comparison between numerical results (solid lines) and experimental data (broken lines) for the time histories of free surface fluctuations at wave gauge #8 ~ #13; $\delta = -0.1$ m, $\gamma = 2.64$. The coordinates for gauges are: Gauge #8: (x, y) = (0.4826, 1.092); Gauge #9: (x, y) = (0.8636, 1.092); Gauge #10: (x, y) = (1.2446, 1.092); Gauge #11: (x, y) = (0.635, 0.4826); Gauge #12: (x, y) = (0.635, 0.8636); Gauge #13: (x, y) = (0.635, 1.2446). The unit is in meter.

The comparisons show that before time = 3 sec, the numerical solution has a good agreement with the laboratory data. After time = 3 sec, the solution and data are contaminated by the reflected wave from the side walls, and will be excluded from current discussion.

5.2.4. Snapshots

We present the snapshots of free-surface elevation, velocities, and shoreline movements from Figs. 5.5–5.7. The basic physical phenomena are very similar to those in the subaerial cases. In Fig. 5.5, we can see that the free surface has smaller free-surface displacement than that in the subaerial one. This is because that the submerged landslide has a lower initial slide elevation and a smaller specific weight.

Figure 5.6 show the velocities at the centerline cross-section. Compared to the subaerial case, the main processes to generate waves and velocities are similar. One important observation is that the velocity on top of the slide has a speed faster than that in the moving slide. This is caused by the three-dimensional effect. We have found that the moving slide generates a three-dimensional eddy, and this eddy transfers the momentum to the centerline location and makes the high velocity jet on top of the slide. This phenomenon can be clearly seen in Fig. 5.6.

Figure 5.7 shows the snapshots of the shorelines. Because of the slide movement, from time = 0.0 sec to 0.9 sec, the runup height has negative values. After time = 1.2 sec, a rebounding wave causes the shoreline to rise up. The maximum runup at the centerline happens at time = 1.75 sec. Compared to the subaerial case (Fig. 4.9, time = 2.7 sec), different shoreline curves can be observed. In the subaerial simulation, the shoreline curve looks like a hump. However, in this subaerial case, the shoreline curve is more like a concave.

5.2.5. Cross-Sections

In this section, the detailed velocity fields (time = 1.5 sec) at different cross-sections will be presented from Figs. 5.8–5.10. Because the major observations are close to that of the subaerial case, only the differences and important phenomena will be addressed.

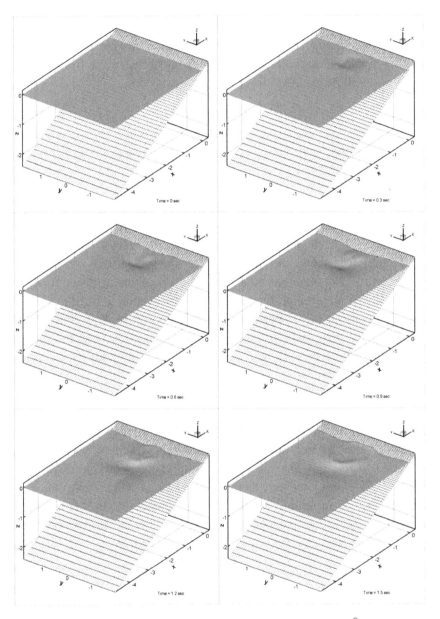

Fig. 5.5. Snapshots of free-surface profile for the sliding wedge with $\delta = -0.1$ m, and $\gamma = 2.64$. The unit is in meter.

Eddy Simulation Model for Tsunami and Runup Generated by Landslides 153

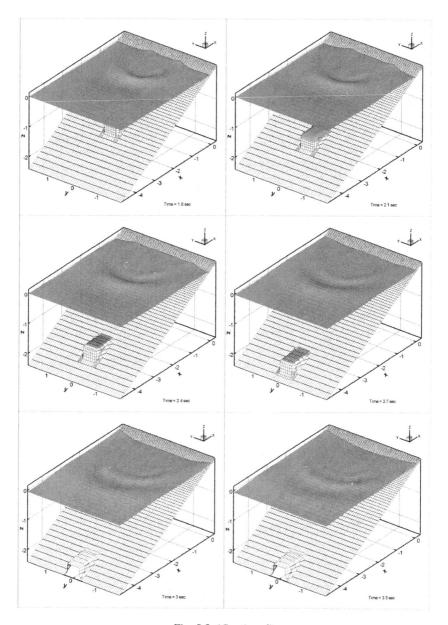

Fig. 5.5. (*Continued*)

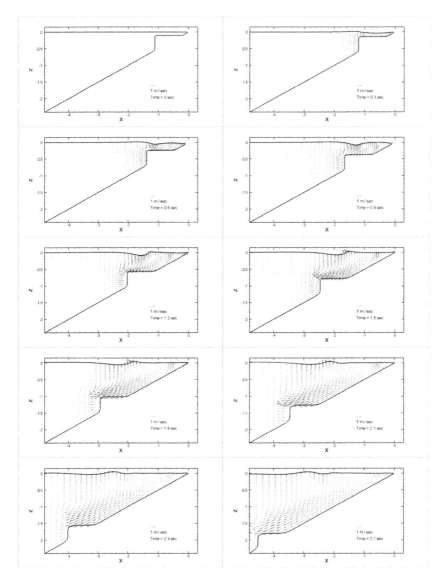

Fig. 5.6. Snapshots of velocity vectors on the centerline vertical plane for the sliding wedge with $\delta = -0.1$ m, and $\gamma = 2.64$. The unit is in meter.

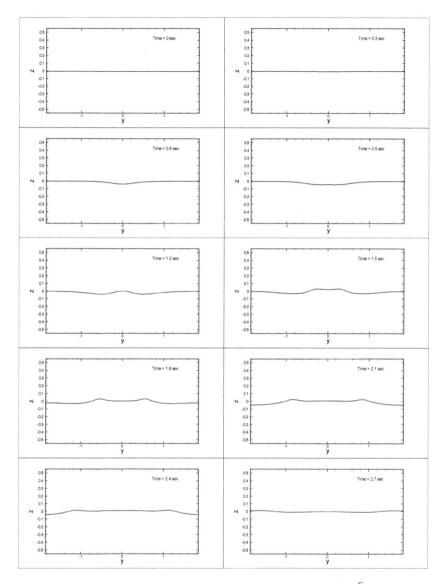

Fig. 5.7. Snapshots of shoreline movement for the sliding wedge with $\delta = -0.1$ m, and $\gamma = 2.64$. The unit is in meter.

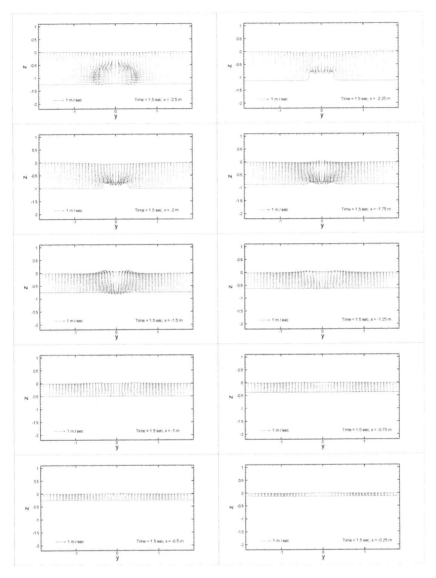

Fig. 5.8. Velocity vectors on the vertical x-planes at time = 1.5 sec for the sliding wedge with $\delta = -0.1$ m, and $\gamma = 2.64$. The unit is in meter. The magnitude of the reference vector indicates the speed of the wedge.

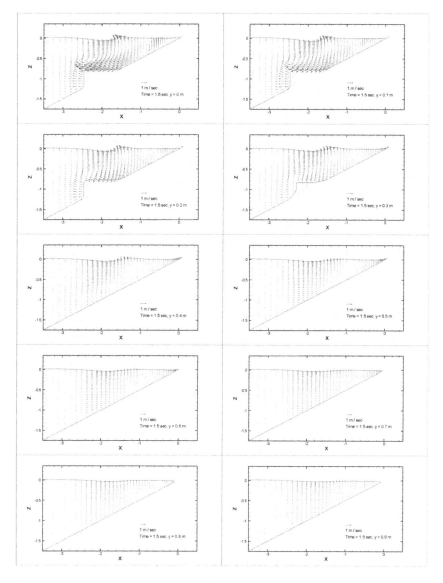

Fig. 5.9. Velocity vectors on the vertical y-planes at time = 1.5 sec for the sliding wedge with $\delta = -0.1$ m, and $\gamma = 2.64$. The unit is in meter. The magnitude of the reference vector indicates the speed of the wedge.

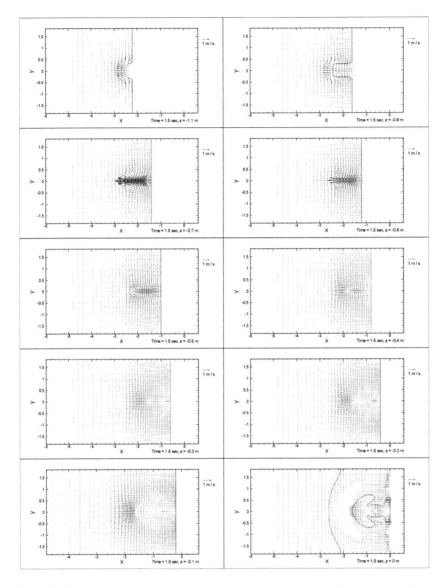

Fig. 5.10. Velocity vectors on the horizontal z-planes at time = 1.5 sec for the sliding wedge with $\delta = -0.1$ m, and $\gamma = 2.64$. The unit is in meter. The magnitude of the reference vector indicates the speed of the wedge.

From Fig. 5.8, $x = -2.0$ m ~ -1.25 m, the three-dimensional effect which makes the momentum concentrate into the centerline can be observed. The velocities on the centerline cross-section are higher than those in the off-center cross-section (Fig. 5.9, $y = 0.0$ m and $y = 0.2$ m). The three-dimensional movement can be also observed from Fig. 5.10, at $z = -0.7$ m and $z = -0.6$ m.

6. Concluding Remarks

In this paper we have presented a LES model for 3D free surface flows. Although the applications shown in the paper are for landslide generated waves and shoreline movements, the model can be used as a tool for studying other practical coastal problems. The advantage of the numerical simulation approach is its ability in providing a large amount of information in the entire flow domain of interest, which is not possible for most of laboratory observations. This is especially true for a large scale experiment.

The success in developing a moving solid algorithm greatly extends the model capability to simulate wave-structure interaction problems. However, further improvement on the turbulent closure modeling for flow separation and wave breaking is essential.

Acknowledgment

This work was supported by research grants from National Science Foundation to Cornell University.

References

1. Bardet, J.-P., Synolakis, C. E., Davies, H. L., Imamura, F., and Okal, E. A. (2003) Landslide Tsunamis: Recent Findings and Research Directions. *Pure Appl. Geohys.* **160**, 1793–1809.
2. Barth, T. J. (1995) Aspects of unstructured grids and finite-volume solvers for Euler and Navier Stokes equations, VKI/NASA/AGARD Special Course on Unstructured Grid Methods for Advection Dominated Flows, AGARD Publication R-787.

3. Bussmann, M. Kothe, D. B., and Sicilian J. M. (2002) Modeling high density ratio incompressible interfacial flows. *Proceedings of FRDSM'02, 2002 ASME Fluids Engineering Division Summer Meeting*. FEDSM2002-31123, 1–7.
4. Cabot, W. and Moin, P. (2000) Approximate wall boundary conditions in the large-eddy simulation of high Reynolds number flow. *Flow Turb. Combust.* **63**, 269–291.
5. Chorin, A. J. (1968) Numerical solution of the Navier-Stokes equations. *Math. Comp.*, **22**, 745–762.
6. Christensen, E. D., and Deigaard R. (2001) Large eddy simulation of breaking waves, *Coast. Eng.* **42**, 53–86.
7. Cox, D. C. and Morgan, J. (1977) Local tsunamis and possible local tsunamis in Hawaii, *HIG 77-14*, Hawaii Institute of Geophysics, University of Hawaii, pp. 118.
8. Deardorff, J. W. (1970) A numerical study of three-dimensional turbulent channel flow at large Reynolds numbers. *J. Fluid Mech.*, **41**(2), 452–480.
9. Dommermuth, D. G., Rottman, J. W., Innis, G. E., and Novikov, E. A. (2002) Numerical simulation of the wake of a towed sphere in a weakly stratified fluid, *J. Fluid Mech.* **473**, 83–101.
10. Fadlun, E. A., Verzicco, R., Orlandi, P., and Mohd-Yusof, J. (2000) Combined immersed-boundary finite-difference methods for three-dimensional complex flow simulations. *J. Comp. Phys.* **161**, 35–60.
11. Goldstein, D., Handler, R., and Sirovich, L. (1993) Modeling a no-slip flow boundary with an external force field, *J. Com. Phys.* **105**, 354.
12. Golub, G. H. and Van Loan, C. F. (1989) *Matrix Computations*. Johns Hopkins University Press.
13. Grilli, S. T., Vogelmann, S., and Watts, P. (2002) Development of a 3D numerical wave tank for modeling tsunami generation by underwater landslides, *Eng Analysis Boundary Elemt.* **26**(4), 301–313.
14. Harlow, F. H. and Welch, J. E. (1965) Numerical calculation of time-dependent viscous incompressible flow. *Phys. Fluids*, **8**, 2182–2189.
15. Heinrich, P. (1992) Nonlinear water waves generated by submarine and aerial landslides. *J. Waterw., Port, Coast., Ocean Eng. ASCE*, **118**(3), 249–266.
16. Hendrikson, K., Shen, L., Yue, D. K. P., Dommermuth, D. G., and Adams, P. (2003) Simulation of steep breaking waves and spray sheets around a ship: The last frontier in computational ship hydrodynamics. *The Resource Newsletter*, U.S. Army Engineer Research and Development Center Information Technology Laboratory, Fall, 2–7.
17. Hirt, C. W. and Nichols, B. D. (1981) Volume of fluid (VOF) method for the dynamics of free boundaries. *J. Comp. Phys.*, **39**, 201–225.
18. Issa, R. I. (1986) Solution of implicitly discretized fluid flow equations by operator-splitting. *J. Comput. Phys.*, **62**, 40–65.
19. Kawata, Y., Benson, B. C., Borrero, J. C., Borrero, J. L., Davies, H. L., DE Lange, W. P., Imamura, F., Letz, H., Nott, J., and Synolakis, C. E. (1999) Tsunami in

Papua New Guinea was as Intense as First Thought, *EOS. Trans. Am. Geophys. Union 80*, **101**, 104–105.
20. Jiang, L., and LeBlond, P. H. (1994) Three-dimensional modeling of tsunami generation due to a submarine landslide, *J. Phys. Oceanog.*, **24**, 559–572.
21. Kim, J., Moin P., and Moser R. (1987) Turbulence statistics in fully developed channel flow at low Reynolds number. *J. Fluid Mech.*, **177**, 133–166.
22. Kothe, D. B., Williams, M. W., Lam K. L., Korzewa, D. R., Tubesing, P. K., and Puckett E. G. (1999) A Second-order accurate, linearity-preserving volume tracking algorithm, for free surface flows on 3-D unstructured meshes, *Proceedings of the 3rd ASME/JSME Joint Fluids Engineering Conference*. FEDSM99-7109, July 18–22.
23. Leonard, A. (1974) Energy cascade in large eddy simulation of turbulent fluid flow. *Adv. Feopgys.* **18A**, 237–248.
24. Li, C.-W. and Lin, P. (2001) A numerical study of three-dimensional wave interaction with a square cylinder. *Ocean Eng.*, **28**, 1545–1555.
25. Lilly, D. K. (1967) The representation of small-scale turbulence in numerical simulation experiments. In H. H. Goldstine (Ed.), *Proc. IBM Scientific Computing Symposium on Environmental Sciences*, IBM Form No. 320-1951, Yorktown Heights, New York, pp. 195–210.
26. Lin, P. and Li, C. W. (2002) A σ-coordinate three-dimensional numerical model for surface wave propagation. *Int. J. Numer. Meth. Fluids*, **38**: 1048–1068.
27. Lin, P. and Li, C. W. (2003) Wave-current interaction with a vertical square cylinder. *Ocean. Eng.*, **30**, 855–876.
28. Lin, P. and Liu, P. L.-F. (1998a) A numerical study of breaking waves in the surf zone. *J. Fluid Mech.*, **359**, 239–264.
29. Lin, P. and Liu, P. L.-F. (1998b) Turbulence transport, vortices dynamics, and solute mixing under plunging breaking waves in surf zone. *J. Geophys. Res.*, **103**(C8), 15677–15694.
30. Liu, P. L.-F., Lin, P. Z., and Chang, K. A. (2001) Numerical modeling of wave interaction with porous structures – Closure, *J. Waterw., Port, Coast., and Ocean Eng., ASCE* **127**(2): 124–124.
31. Liu, P. L.-F., Lin P. Z. and Chang, K. A. (1999) Numerical modeling of wave interaction with porous structures, *J. Waterw., Port, Coast., and Ocean Eng., ASCE* **125**(6): 322–330.
32. Liu, P. L.-F., Wu, T.-R., Raichlen, F., Synolakis, C. E., and Borrero, J. C. (2005) Runup and rundown generated by three-dimensional sliding masses, *J. Fluid Mechanics* **536**: 107–144.
33. Lynett, P., Wu, T.-R. and Lim P. L.-F. (2002) Modeling Wave Runup with Depth-Integrated Equations, *Coastal Engineering*, **46**(2), 89–107.
34. Mohd-Yusof, J. (1997) Combined immersed boundaries/B-splines methods for simulations of flows in complex geometries, CTR Annual Research Briefs, NASA Ames/Stanford University.

35. Pope, S. B., *Turbulent Flows* (Cambridge University Press, 2001).
36. Press, W. H., Teukolsky, S. A., Vetterling, W. T., and Flannery, B. P., Numerical recipes in Fortran (Cambridge, 1986).
37. Raichlen, F. and Synolakis, C. E. (2003) Run-up from three dimensional sliding mass, in *Proceedings of the Long Wave Symposium 2003*, (eds. Briggs, M, Coutitas, Ch.) XXX IAHR Congress Proceedings, ISBN-960-243-593-3, 247–256.
38. Rider, W. J. and Kothe, D. B. (1998) Reconstructing Volume Tracking, *J. Comp. Phys.*, **141**, 112–152.
39. Shen, L. and Yue, D. K. P. (2001) Large-eddy simulation of free-surface turbulences, *J. Fluid Mech.*, **440**, 75–116.
40. Smagorinsky, J. (1963) General circulation experiments with the primitive equations: I. The basic equations, *Mon. Weather Rev.* **91**, 99–164.
41. Synolakis, C. E., Bardet, J. P., Borrero, J. C., Davies, H., Okal, E. O., Silver, E. A., Sweet, S., and Tappin, D. R. (2002) Slump origin of the 1998 Papua New Guinea Tsunami, *Proc. Royal Society, London, Ser. A*, **458**, 763–789.
42. Synolakis, C. E., Raichlen, F. (2003) Waves and runup generated by a three-dimensional sliding mass, in *Submarine mass movements and their consequences*, Advances in Natural Hazards, **19**, eds. Locat, J. and Mienert, J. (Kluwer Academic Publishers, Dordrect, 113–120).
43. Varga, R., Matrix Iterative Analysis (Prentice-Hall, Englewood Cliffs, NJ, 1962).
44. Verzicco, R., Mothod-Yusof, J., and Orlandi, P. (2000) Large eddy simulation in complex geometric configurations using boundary body forces, *AIAA Journal* **38**(3), 427–433.
45. Watanabe, Y. and Saeki, H. (1999) Three-dimensional large eddy simulation of breaking waves, *Coast. Eng.*, **41**(3, 4), 281–301.

CHAPTER 5

FREE-SURFACE LATTICE BOLTZMANN MODELING IN SINGLE PHASE FLOWS

J. B. Frandsen

School of Civil Engineering
The University of Sydney, NSW 2006, Australia
Email: jbehrndtz@yahoo.com

The mathematical framework of kinetic theory allows for investigations of micro structures in the flow at the mesoscopic level and thus, in theory, offers further insight into underlying mechanisms of nonlinear processes at the free surface than traditional Navier-Stokes and Non-Linear Shallow Water (NLSW) solvers. This is of course only true if the lattice spacing is sufficiently small. The Lattice Boltzmann (LB) method simulates fluid flow by tracking particle distributions in a Lagrangian manner. The particles are constrained to move on lattices. The Boltzmann equation relates the time evolution and spatial variation of a collection of molecules to a collision operator that describes the interaction of the molecules. Mathematically, the collision integral of the LB Equation (LBE) poses difficulties when solution of the equation is sought. Investigators overcome this through descriptions of models with different levels of accuracy in the approximations of the integral. We consider a model in which the collision assumptions are simplified to a single-time relaxation form. This is the simplest form of the LBE. It is referred to as the Lattice Bhatnagar-Gross-Krook (LBGK) scheme. In the present chapter, we present solutions primarily based on the standard LBE which herein approximate the NLSW equations in rotational flow. We show some preliminary LBGK test cases including weak bore and dam break predictions. Fairly good agreement was found with Riemann solutions and model scale experiments. With reference to the test cases of the present workshop, the tsunami-generated runup onto a plane beach (benchmark 1) was also simulated; presented in Chapter 16 in this volume. We tested the standard LBE scheme and a second-order Finite-Difference (FD) LB model. The LB schemes with shoreline algorithm of Lynett *et al.* (2002) worked well whereas the thin film approach would result in inaccurate velocities. We highly recommend the second-order FD LB scheme to form a basis for free-surface water wave developments.

1. Introduction and Overview

There are several kinetic or mesoscopic methods, e.g. the lattice gas cellular, the lattice Boltzmann equation, the gas kinetic schemes, the smoothed particle hydrodynamics, and the dissipative particle dynamics method. The Lattice Boltzmann (LB) method with roots in the Boltzmann equation stems from the ideas of Ludwig Boltzmann (1844-1906) who made advances in electromagnetism and thermodynamics. We seek to utilize his gas dynamics theory (Boltzmann, 1964). Lattice gas models and LB models have been used in fluid mechanics since the 1990s. About 1000 articles have been published over the last fourteen years. The majority of these publications are related to physics and computer sciences (e.g. soft condensed matter, porous media, microfluidics, electrokinetic flows, magnetohydrodynamics, etc.). The applications are broad. Sukop and Thorne (2006) have provided an overview of a number of papers published (1992-2004), divided into various areas of science, in the introductory chapter of their recent book.

The purpose of this chapter is to introduce the LB Equations (LBE), the simplified Boltzmann Equation, with application to free-surface water waves. Applying the LB approach will allow us to solve the fluid flow at the mesoscopic scale level, i.e. length scales ranging between the atomistic or micro and the continuum mechanics level (assuming the grid refinement is adequate). The latter is hereafter referred to as the macro scale level. Somewhere between the atomistic and macro scale level (Fig. 1(a)), the continuum model approach breaks down where part of the physical systems cannot be assumed to be continuous (e.g. breaking waves). This is one of the main motivations behind exploring the LB approach. Herein, we shall, however, focus on single phase LB models only, as this is our first experience with LB modeling. We have highlighted some questions which we asked ourselves before we chose to explore the LB approach. Some of these modeling features and issues are shown in Fig. 1(b). We have further indicated our progress on what we have/have not tested. The first important question to ask was whether or not we could predict nonlinearities at the free-surface. We have got some promising indication that the LB method may work, as shown in the test case studies in §6. Our investigations have so far been limited to shallow water depths. We do not yet know how accurate the method would be regarding wave breaking predictions. One feature which motivated us to chose the LB approach is that we can model bubble break-up. Regarding Fluid-Structure interaction (FSI) we have so far

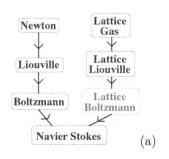

Topic in question	LBM
Grid dependent soln.	yes
Wave continuous	yes
Nonlinear waves	yes
Free-Surface Algorithm	yes/no
Wave breaking	?
Bubble break-up	yes
Poisson freedom	yes
Viscosity	water/challenge
BC simple	yes
FSI /arbitrary geometry/topologies	yes/no
Efficient parallelization	yes/no
CPU efficiency	ok/so far

Fig. 1. Description of Fluid Motion. (a) "Length scale" hierarchy. The microscopic details reduce when moving toward the Navier-Stokes equation. (b) Lattice Boltzmann Model (LBM) properties.

found that the LB approach can provide a relatively easy means of simulating flow around single and multiple bluff bodies. The standard LB method is based on uniform lattices. This presents the usual difficulties as to grid refinement requirement, and thus yields a host of questions on accuracy and efficiency. This is on-going research, as described in the review part of this chapter.

Further, we should point out that our intention with the LB modeling approach is to use it in local areas where physical details are needed such as the nearshore regions and in the breaking wave process itself. Therefore a realistic goal and potential improvement to the current tsunami model literature would be to propose an LB model coupled with a macro-scale level model for large domain tsunami model predictions.

The key issues to bear in mind when reading through this chapter are that the standard LBE scheme/approach has the following model properties:

- the fluid mass is collected in discrete particles, located on corner points of the lattice site;
- the velocity in a lattice node is discretized (in addition to time and space);
- the particles move according to a finite, discrete set of velocities;
- the viscosity (ν) is introduced via the single time relaxation parameter (τ) and is a discretized equation with uniform lattice spacing Δx and time step Δt, $\nu = \frac{c^2}{3}(\tau - \frac{\Delta t}{2})$ where $c = \Delta x/\Delta t$. In the incompressible flow simulation herein, we must ensure that ν is constant;

- the formulation contains a linear convection operator;
- the solutions to the Navier-Stokes equations are in the nearly incompressible limit;
- the solutions herein are based on the depth integrated shallow water equations in rotational flows;
- the solutions to the LBE can be shown to satisfy the Navier-Stokes equations (§4.4);
- the fluid pressures p are calculated from the equation of state: $p = c_s^2 \rho$ where $c_s = \sqrt{\partial p/\partial \rho}$ ($= c/\sqrt{3}$) is the speed of sound and ρ is the fluid density;
- the transformation between the microscopic level to the macroscopic variables consists of simple relationships;
- the standard LBE is second-order accurate in space and first-order in time (§4.6);
- the fundamental differential equations of gas dynamics are similar to those of shallow water wave theory.

From our viewpoint, the LB model is mainly proposed due to the molecular properties of the formulation and its ease of handling large data sets in relation to understanding underlying mechanisms of breaking ocean waves. As outlined above, there are of course several other important reasons an LB modeling approach appears to be a promising simulation method in fluid dynamics. Some of the numerical model advantages of the LB method are: (1) no Poisson equation, and (2) a reduction from second-order to first-order partial differential equations, and therefore a simplification of nonlinear free-surface numerical modeling. However, as mentioned, the major motivation of the present work is driven by the mathematical framework allowing molecular level modeling of fluid flow which differs from any conventional free-surface models. The locality of the formulation is another advantage, essential for CPU efficiency and thus practicality. As for most fluid dynamics problems, high CPU requirement is a barrier in advancing free-surface models which typically involve high CPU time near the moving boundary, especially when discontinuities occur. The LB method has been demonstrated to show good parallel computing performance (e.g. Aidun *et al.* (1998); Thürey *et al.* (2006a); Körner *et al.* (2006)). Furthermore, compared to traditional continuum mechanics solvers, the LB solver has also shown promise when simulating fluid interaction with multiple bluff-bodies in terms of low CPU time and an easy means of rearranging bluff-bodies to study optimum solutions for flow-induced vibration problems with/without a free surface. An example of bore propagation through a coastal city is shown in Fig. 2.

Fig. 2. Bore propagation through a city.

The standard LB method is based on uniform grids. Although it is an advantage that no conventional mesh generation is required, it is usually inadequate for most fluid dynamics problems. Compared to traditional CFD, few initiatives have been done on unstructured LB lattices, as mentioned later in the review.

In this chapter, we have assumed that ocean and coastal engineers are not familiar with LB-based models and therefore, this chapter is designed to be as self-contained as possible. Having said this, the reader is encouraged to carry out a self-study on physical gas dynamics to understand basic kinetic theory and the core idea of connecting microscopic and macroscopic physics before reading about Boltzmann modeling concepts, e.g. Vincenti and Kruger, Jr. (1965); Burgers (1969); Gombosi (1994); Laney (1998); Struchtrup (2005). It is also recommended to acquire some knowledge in statistical mechanics, e.g. Penrose (1970); Kalikmanov (2001). Among other useful books in the field of LB Equations and modeling are those of Cercignani (1988); Succi (2001); Karniadakis and Beskok (2002); Harris (2004); Zhou (2004); Sukop and Thorne (2006). We should also point out that the LBE was developed in the wake of Lattice Gas Cellular Automata (LGCA). The basic ideas of LGCA have been extensively described by, for example, Rothman and Zaleski (1997); Rivet and Boon (2001).

The first part of the present chapter starts with a short review of LB models. The description of fluid motion is divided into an introduction to kinetic theory followed by the Boltzmann equation. Then a presentation of the approximate form of the collision integral is given; leading into the numerical treatment. The second part of the chapter focuses on the LB development with application to free-surface water waves. The solver performance is first tested for weak bores in tanks, then a dam-break problem and finally the tsunami runup simulation is undertaken (presented in this chapter). We should point out that the case studies presented represent preliminary work. Further work is required to address moving shore line dynamics, etc., in a well documented manner.

2. Review of Lattice Boltzmann Models

This brief review is meant to provide the reader with insight into the work and progress on development and applications of LBE, in particular, with flow around bluff-bodies and free-surface flows in mind. We emphasize that this approach has not been applied and validated in the area of free-surface ocean wave predictions. Therefore, a review in a more general format is presented. We outline the development of various branches of LB modeling. In doing so we highlight ideas of expansion and development of LB-based approaches relevant to the ocean and coastal engineering communities.

The LB formulation offers an alternative treatment for description of fluid particle motion. It is based on statistical mechanics concepts and recovers the Navier-Stokes equations in the nearly incompressible limit, as mentioned by Chen and Doolen (1998). The approach has been used extensively in molecular dynamics and physics but little attention has been given to the modeling of free-surface water waves. The LB method was developed as an improvement of the method of Lattice Gas Automata (LGA), as described by Rothman and Zaleski (1997). It was first introduced by McNamara and Zanetti (1988). Comprehensive reviews of the LB method have been given (Benzi et al. (1992); Chen and Doolen (1998); Nourgaliev et al. (2003)). The LB method belongs to a class of the pseudocompressible solvers of the Navier-Stokes equations and can be classified as a Lagrangian, local equilibrium, finite-hyperbolicity approximation. The numerical formulation typically (but not necessarily) separates the equations into the dissipative and non-dissipative parts. The most popular form of LB equations is the Lattice-BGK (LBGK) incorporating a single-time approximation of the Boltzmann equation, as described by Bhatnagar et al. (1954), hereafter the LBGK model. In this model the mean free path between particles is assumed to be the same. It should also be noted that the governing LB formulation is second-order accurate space and first-order accurate in time, as shown by Junk et al. (2005). The disadvantages inherent in the LBGK model can be reduced by using a Multi-Relaxation-Time (MRT) approach. It separates the relaxation times for different kinetic modes and improves numerical stability and accuracy. For more detailed discussions on MRT-LBE method in general, we refer the reader to D'Humières et al. (2002). It is notable that the trend in progress of LB modeling is reported in relation to the MRT approach. Stability of the LBE method is further addressed through the generalized Lattice Boltzmann equation which is based on moment space rather than discrete velocity space, e.g. Lallemand and Luo (2000); Luo (2000).

2.1. Grid Refinement and Curved Solid Boundaries

The majority of LBE model predictions are based on uniform regular spaced lattices. Compared to traditional CFD, few initiatives have been done on unstructured LB lattices. Although an increase of contributions is evident, a literature review still reveals only relatively few publications on this topic, e.g. He et al. (1996); Tölke et al. (1998); Peng et al. (1999); Kandhai et al. (2000); Van der Sman and Ernst (2000). Mesh refinement techniques are also discussed by Crouse et al. (2002); Ubertini et al. (2003); Crouse et al. (2003). Recent progress on grid refinement method using a nested adaptive grid approach are described by Tölke et al. (2006), in which numerical examples with rising bubble simulations are presented. Geller et al. (2006) simulates flow around multiple cylinders based on unstructured grids at $Re=200$. Good agreement was found with traditional CFD solvers. Several investigators have made contributions regarding the treatment of curved boundaries on uniform grids in 2-D, e.g. Filippova and Hänel (1997); Mei et al. (1999). Expansion into 3-D flows have also been developed, as described by Mei et al. (2000). We should also mention that "stair-step" effects on the boundary can be overcome by using mapping techniques, like in traditional finite difference models. For example, He and Doolen (1997) developed a curvilinear formulation to simulate flow past a circular cylinder.

Today, the research group directed by Professor Krafczyk at the Institute for Computer Applications in Civil Engineering, Technical University of Braunschweig, Germany, is at the forefront on LB-grid method development. We should also mention that they are in the forefront in many other LB developments as well, including contributions to civil/mechanical engineering applications. The reader is invited to visit their web site which has many publications available for downloading including a nice article search engine (www.cab.bau.tu-bs.de).

With respect to LB-model development with application to steep and breaking free-surface water waves, it would be necessary to have models with non-uniform grids, just like it is when using traditional CFD finite volume and finite element solvers. In contrast to the macro mechanics models, it should be noted that non-uniform grid implementation in the LB context is non-trivial.

2.2. Turbulence Modeling

Capturing turbulence accurately and identification of appropriate subgrid scale laws are topics of research in the LB community. From our view point, one central idea behind proposing the mesoscopic LB formulation is to capture smaller scales naturally, and therefore postpone/avoid the need to apply empirical turbulence models. Of course this highly depends on grid resolution and on the application in mind. However, it may be needed in civil engineering applications with high Reynolds numbers ($Re > 10^6$) or high Froude numbers ($Fr > 1$). Note that the reader will often in the fundamental LBE literature come across the mention of "high" Re in the order of one. Indeed, this is very misleading for researchers involved with civil engineering hydro-and aerodynamics applications. Although turbulence modeling in a Lattice Boltzmann framework is a relatively young research field, some research has been carried out. The limitations of an LB approach in the context of turbulence modeling are not well defined since few engineering related studies have been undertaken. Some studies in fundamental theoretical LBE do indicate additional terms needed to capture smaller scales. Having said this, traditional mesoscopic model applications usually involve length scales in the order of 10–100 nm. The investigations undertaken typically involve combining traditional subgrid models with the LB method. Most of the investigations are typically done with a LBGK or MRT-LBE collison model combined with a Smagorinsky model (Lesieur et al. (2005)). The particle distribution function and hydrodynamic moments will subsequently have to be modified from the standard form, Eq. (11), including a total viscosity $\nu = \nu_0 + \nu_{SGS}$, where ν_0 and ν_{SGS} are the basic kinematic viscosity and the eddy viscosity, respectively. Some examples can be found in the book by Zhou (2004) and in the articles by Krafczyk et al. (2003) and Yu et al. (2006). An example is also given by Hou et al. (1996) who developed an LB subgrid model, proposing to incorporate space-filtered particle distribution functions. Hou et al. (1996) present high Re flows $100 < Re < 1 \times 10^6$ for driven cavity flows. Another interesting study was done by Derksen (2005) who compared the performance of different turbulence models including a standard Smagorinsky and a Sagaut mixed-scale model with application to flow in a swirl tube at $Re \approx 2000$. It is notable that Derksen (2005) successfully made use of an immersed boundary method to avoid stair-step surfaces combined with wall damping functions to treat flow at solid curved walls accurately. Krafczyk et al. (2003) undertook MRT-LBE large eddy simulation studies of flow around a cube at $Re = 40{,}000$ using the standard no-slip bounce back rules at boundaries. Direct Numerical Simulations (DNS) of turbulence in the LB framework (the grid Re is less than 1) has also been undertaken, e.g. Luo et al. (2002); Yu et al. (2005); Lee et al. (2006). Multi-block grid refinement for turbulence model predictions were recently developed by Yu and Girimaji (2006).

It should also be mentioned that a 3-D commercial solver, "*PowerFLOW*", exists including subgrid scale modeling capabilities (Chen et al. (2004); Li et al. (2004)). It has 3-D modeling capabilities but without free-surface algorithms. The code is distributed by *Exa Corporation* (www.exa.com) with application to the aerodynamics/automobile industry, allowing the user to model relatively high Re flow past fixed bluff-bodies. Extensive evaluation of the code has been undertaken, as described by, for example, Lockard et al. (2002). Other publications based on the *PowerFLOW* code are listed on the web site of *Exa*.

2.3. Key Contributions

We should mention that the annual meeting, International Conference for Mesoscopic Methods in Engineering and Science (ICMMES), www.icmmes.org, provides an excellent forum of LBM discussions. The meeting is a relatively new effort initiated to bridge the gap between theory and applications. The first ICMMES was held in 2004. The ICMMES proceedings are usually published in two journals: one in statistical and/or computational physics in general, and the other in computational fluid dynamics (CFD). Currently following special issues have been published:

- 2005: Journal of Statistical Physics, Vol. 121(1/2), 2005.
- 2006: Computers and Fluids, Vol. 35 (8/9); Intl. Journal of CFD, Vol. 20(6).
- 2008: Progress in Computational Fluid Dynamics, Vol. 8 (1/2/3/4).

Studying papers in these volumes alone would bring the reader quickly up to date in the area of Lattice Boltzmann modeling (we have referenced several articles herein). We should also mention that fundamentals on LBE (and beyond) can be found in a series of publications and lecture notes by Professor L.-S. Luo and co-workers, Dept. of Mathematics & Statistics, Old Dominion University, USA. These works are available on the web site: www.lions.odu.edu/~lluo. It is notable that most (if not all!) references on fundamental LBE development and progress can be tracked down through the publications of Luo. The same can be said about the research of Professor Krafczyk and co-workers which also has a focus on engineering applications (www.cab.bau.tu-bs.de).

The book of Sukop and Thorne (2006) is also recommendable as an introduction to the LB method. The authors share many LB experiences (including code material!). The code is available online (www.fiu.edu/~sukopm/) amongst many links to LB investigators, lecture notes, books, on-line LB codes, etc.

3. Review of Free-surface Lattice Boltzmann Models

Several research contributions have been helpful in guiding the present work of long water wave runup. Amongst these that can be mentioned are the free-surface and internal wave model development of Salmon (1999a); He *et al.* (1999); Ghidaoui *et al.* (2001); Deng *et al.* (2001); Xu (2002); Ginzburg and Steiner (2003); Buick and Greated (2003); Krafczyk and Tölke (2004); Zhou (2004); Körner *et al.* (2005); Ghidaoui *et al.* (2006); Thürey *et al.* (2006b). We shall highlight some of these free-surface model developments.

3.1. *Non-Overturning Surfaces in Shallow Liquid Depths*

Common for all of these works of kinetic model approximation of the depth-integrated Navier-Stokes equations is that there is no need to formulate a free-surface algorithm when assuming non-overturning waves. The reason is that the form of the kinetic scheme allows one to solve for the variables of the non-homogeneous shallow water equations automatically. The kinetic scheme utilizes the microscopic particle distribution function as the basis to construct the fluxes. Linear and more importantly non-linearities at the surface are introduced and captured through the collision term. Further, it means that the NLSW LB scheme does not include the kinematic and dynamic free-surface boundary prescribed, as usually required and known from traditional free-surface numerical modeling approaches. To understand this further, one first needs to recognize the similarity between the equations for shallow water theory and the fundamental differential equations of gas dynamics (see the references listed in §4.7). We shall later introduce kinetic theory which forms the background of LB models (§4.1).

In the context of kinetic modeling of water waves, the approximation of the collision integral in the Boltzmann equation becomes very important. We should say that the LBGK approximation is the crudest approximation of the possibilities the LB model series would offer. Since, the LB free-surface water wave research is not an established area, the LBGK approach, the simplest form, would however be a natural starting point for model development. As far as we know, there exists no attempt yet which includes a LB Navier-Stokes formulation for non-overturning, as well as breaking waves, in shallow water depth which has undergone validation (the same can be said for LB deep water predictions which also has not yet been explored). We believe that this has a simple explanation, that is, LB investigators come with very different backgrounds and interests, not to mention the numerical approach is a relative new research area.

The first attempt on LB non-overturning free-surface flows was undertaken by Salmon (1999a). The single phase LBGK model of Salmon obeys the shallow water equations in rotating flows. Salmon's interest was to use the LB model for 3-D ocean circulation models [Salmon (1999b)]. Therefore, a relatively large flow domain in the order of 4000×4000 km^2 was prescribed. It is also notable that the numerical results were based on relatively coarse models of 100×100 lattice points. Dellar (2002) undertakes a critical review of the stability of the shallow water LBE. Zhong et al. (2005) is also interested in a large scale rotational flow field with application to ocean circulation. They expanded the model of Salmon (1999a) to include a fully explicit second-order accurate model. Ghidaoui et al. (2001) and Deng et al. (2001) developed a finite volume BGK model for open channel flows and contaminant flows, respectively. Xu (2002) investigated the performance of a BGK scheme, especially the treatment of shocks. Similar to the formulation of Salmon (1999a), Zhou (2002) explored LBGK solutions including an eddy viscosity model which represented shallow water equations with rotating flows. Zhou (2004) accumulated many of his investigations in book format with a focus on flow effects in channels on rough beds with application to river mechanics problems.

Recent progress in development and applications are by Que and Xu (2006); Ghidaoui et al. (2006); Frandsen (2006a); Thömmes et al. (2007). Thömmes et al. (2007) applied their LBGK solver to complex geometries with irregular bathymetry with application to flow in the Strait of Gibraltar. From the view point of river mechanics and bathymetry implementation, it is encouraging to learn about LB model usages. However, the free-surface solutions were somewhat non-physical; some results are also similar to Zhou (2004). Ghidaoui et al. (2006) have developed a Finite Volume (FV) BGK solver for shallow water flows around circular cylinders. They reported on wake flow physics for $Re < 200,000$. Que and Xu (2006) also developed FV BGK to model roll and solitary waves. Relevant to the present workshop, they simulated runup. To handle the wet-dry interface, they prescribed a thin film layer on the beach slope. Comparison was done with the experimental data of Synolakis (1987). Reasonable agreements were found in terms of the free-surface elevations. However, the velocity components were not reported on. We found that this variable drives the resolution required. In other words, we found that while the free-surface elevation would compare well with experimental data, this would not mean that the velocity components would. See further details on this in benchmark test case 1, as reported in Chapter 16 in this volume.

Common to the above publication is the missing adequate time evolutions of the free-surface elevation and velocities. Without report on the time histories of the important variables, one cannot make any judgement as to whether or not a numerical free-surface solver performs adequately. We initiated some simulations based on a standard LBGK solver (Frandsen (2006a)). Promising free-surface results were found in weak bore formation studies. The weak jump in the surface agreed fairly well with the physical model scale tests. The single phase LB model has demonstrated promise and competitive performance compared to a Riemann solver and other free-surface solutions. However, these solutions may be viewed as finite difference discretizations of the depth-averaged Navier-Stokes equations, and not mesoscopic level flow solutions.

3.2. *Breaking Surfaces in Any Liquid Depths*

Common for all of these works is that a free-surface algorithm is embedded in the LB kinetic frame work. Although we have not seen LB literature for linear and steep waves free-surface algorithm, there are developments of LB models with allowance for break-up of the free-surface. Ginzburg and Steiner (2003) developed an immiscible LB free-surface model in single phase flow. They proposed to treat the LB free-surface with a Volume of Fluid (VOF) approach within the framework of the generalized LBE. The test cases on regular grids had application to filling processes in casting and were treated as one fluid (air particle interactions on the melt metal were ignored) in 2-D and 3-D predictions. Filling simulations included dense fluids of Reynolds no. <717 and Froude no. <10.7. The free-surface solutions were based on a first-order Chapman-Enskog expansion of the distribution functions and a first-order explicit upwind scheme. Undesirable fluctuations were dampened by inclusion of a Smagorinsky model. Since the free-surface model of Ginzburg and Steiner (2003) was approximated using the VOF type of approach at the free-surface, it should in principle work for two phase flows. However, this remains to be explored and how well this approach works for LB approximate water wave behavior remains unanswered too. Obviously liquid with greater viscosity than water exhibits quite different behavior and is in some sense easier to model, as the free-surface motion is dampened. Later, Körner et al. (2005) developed a laminar free-surface LBGK model with application to foaming processes, especially foaming of metal. The free-surface was treated with a VOF type of approach and regarded as a surface capturing method (Eulerian). Their bubble simulation results covered 2-D advection-diffusion with D_2Q_4 velocity sets. Krafczyk and Tölke (2004) extended the model of Körner et al. (2005) by including a large eddy simulation model to describe turbulent flows. Their test cases included an adaptive 3D simulation of the classical dam-break problem and interactions with various structures. The most recent research is undertaken by Thürey (2006) who has developed various free-surface LB solvers, allowing for wave breaking based on VOF (Thürey et al. (2006)) and level set algorithms (Thürey and Rüde (2004)). Thürey recommends the VOF algorithm due to the CPU time efficiency. Solvers of Thürey's includes the discretization the NLSW equations and the Navier-Stokes equations, respectively. Thürey et al. (2006b) present VOF simulations with application to ship hydrodynamics, based on an LB solver which couples a 2-D shallow water formulation, after Zhou (2004), with 3-D Navier-Stokes flow. It should be mentioned that the research of Thürey has a focus on computer graphic development. The visual effects shown in images and animations are very impressive (www.ntoken.com). Some source code is also made available (elbeem.sourceforge.net/).

3.3. Bubbles and Drops

Most LB publications assume that the molecules behave as hard spheres and collide elastically. However, we should mention that a strength of the

LB model approach and our attraction to this approach is that one can take the collision approach much further to include soft sphere modeling, as highlighted in Fig. 1(b). For example, Professor Yeoman and co-workers (Oxford, Theoretical Physics) have undertaken work on this aspect of LB modeling in the general area of binary fluids, wetting/spreading, and the application of droplet impact predictions (funded by the inkjet/laserjet printer industry). The group has also made advances made in liquid crystal dynamics, e.g. Denniston et al. (2004). Abraham and co-workers (Purdue University, Mechanical Engineering) have conducted a variety of bubble interaction dynamics experiments, e.g. McCracken (2004); Premnath (2004); Premnath and Abraham (2005, 2007). These experiences are useful starting points for the free-surface breaking-wave application. Furthermore, Professor Aidun and co-workers (Georgia Tech, Mechanical Engineering) have simulated stokes flow with deformable particles amongst a long list of contributions in the area of suspended particles hydrodynamics, e.g. Ding and Aidun (2003). Some of the fundamental research have been applied to biomedical problems, e.g. Ding and Aidun (2006). Recently, Aidun et al. have introduced an LB-fluid model with deformable finite-elements particles to simulate mesoscale blood flow. The work, in preparation for publication, was presented at the 2006 meeting of the American Physical Society, Division of Fluid Dynamics.

Finally we note that it is still an unresolved LB problem as to how to tackle the high density ratio difference at the free-surface interface in two phase flow in ocean and costal engineering applications. Some work on this has been initiated in relation to bubble and droplets, as described by, e.g. Inamuro et al. (2004); Lee and Lin (2005).

These LB experiences mentioned in this section, although not obvious at first glance, are all very useful starting points for the free-surface breaking-wave application in ocean and coastal engineering.

3.4. *Status on LB Free-Surface Water Wave Validation*

In our review on free-surface LB solver development, we did not come across validation material necessary to make judgement as to whether or not an LB approach would be a reliable type of approach to predict linear and/or nonlinear water wave behavior. Recently, we did some bore prediction studies based on a standard LBE solver which included relatively long time evolutions of the free-surface. The time histories of the free-surface compared fairly well with experimental results, as shown in §6.2 and in Frandsen

(2006a). The LBGK solver did not reveal any sign of numerical dissipation but an excessive resolution was required to simulate the weak jump at the surface. We have done some further stability assessment through the wave runup studies herein; see benchmark test case 1 reported in Chapter 16 in this volume. These studies suggest that high-order finite difference LB schemes improve the instability problems typically found in the evolution of the velocity components. Having said this, we cannot do direct comparisons between the bore studies and wave runup problems, as the latter test case is governed and influenced by the treatment of the wet-dry interface.

Our observations based on the current progress do suggest advantages when using high-order finite-difference LB schemes. Some of our LB experiences indicate that this kind of LB model should be a prerequisite for breaking LB wave models. In our free-surface review, it was notable that some investigators have applied the standard LBE with turbulence models and a breaking-wave algorithm (VOF). Some investigators have used the means of turbulence models to suppress instabilities. Some LB solvers have also gone to the extent of tuning the viscosity, i.e. so that the fluid is not a water representation. From a numerical water wave mechanics development view point, none of these are appropriate steps to take. Although the above-mentioned research forms some of the background necessary for the development of the LB free-surface framework, it is still uncertain if an LB approach can simulate water wave behavior accurately. Little research on water waves using LB modeling have been undertaken. The LB research is very much in the beginning of entering the group of numerical solver options available to ocean and coastal ocean engineering applications. On a final note, we conclude that it has not yet been demonstrated that non-overturning water waves based on standard LBE formulations or other LB models are an adequate means of describing free-surface water wave behavior.

4. The Governing Formulation

The present model is based on kinetic theory which attempts to describe the macroscopic fluid behavior using the laws of mechanics and probability theory. Provided that the fluid is near a state of equilibrium and the hydrodynamics moments of the equilibrium distribution functions are conserved, kinetic theory can be shown to satisfy the Navier-Stokes equations.

4.1. *Kinetic Description of Fluid Motion*

The LB Equation (LBE) originates from the kinetic theory of gases and is a minimal form of the Boltzmann kinetic equation in which all the details of molecular motion are removed, except those that are strictly needed to recover hydrodynamic behavior at the macroscopic level such as mass and momentum conservation on sufficient resolution of the lattices (Succi, 2001). The LB method discretizes kinetic theory.

Kinetic theory assumes that the fluid is described by a large number of molecular constituents whose motions obey Newtonian mechanics. The objective is not to know the motion of every individual molecule, but the collective behavior for which we need a statistical description of the system, as illustrated in Fig. 3.

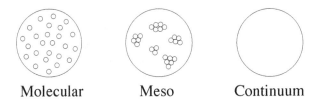

Fig. 3. The LB model mimics collective molecular behavior at the mesoscopic level (i.e. at length scales in between the atomistic and continuum mechanics levels), if the lattice spacing is sufficiently small.

The statistical description of a fluid at or near equilibrium is contained in the single-particle distribution function, $f_i(x, c_i, t)$, where x represents spatial coordinates, c_i represents microscopic velocity of molecules and t denotes time. This function resides in a time-evolving phase space. The goal of kinetic theory is to formulate and solve a transport equation governing the time and spatial evolution of the distribution function with different collision processes dictated by the nature of the interactions between molecules, as described by Gombosi (1994). The flow field is represented by particles which are constraint to move from lattice node to lattice node, i.e. in a Lagrangian manner. The particles naturally stream/propagate and collide when meet. After collision, the particles stream to adjacent lattice nodes. In contrast to continuum based numerical models, where only space and time are discrete, the discrete variables of the LB model are space, time and particle velocity. The fluid mass is collected in discrete

Table 1. Velocity set of the D_2Q_9 lattice.

i	0	1	2	3	4	5	6	7	8
$c_{i,x}$	0	+c	0	-c	0	+c	-c	-c	+c
$c_{i,y}$	0	0	+c	0	-c	+c	+c	-c	-c

(lattice) particles. The particles are by default located at the corner nodes of a regular lattice. The particles move according to a finite, discrete set of velocities. The standard LB model notation follows a D_nQ_m reference where n is dimension and m denotes number of particles. A simple example is the 1-D diffusion model or following the notation a D_1Q_2 lattice model. A 2-D convection-diffusion model would be referred to as a D_2Q_9 lattice model and the 3-D counter part as a D_3Q_{19} lattice model. The velocity components of the D_2Q_9 model are illustrated in Fig. 4.

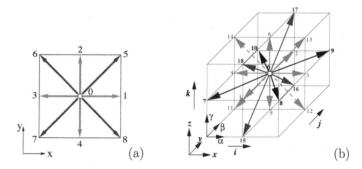

Fig. 4. (a) D_2Q_9 and (b) D_3Q_{19} velocity components.

The propagating direction of the discrete velocities vectors of the D_2Q_9 model is given by

$$c_{i=1-4} = (cos[(i-1)\pi/2], sin[(i-1)\pi/2])c \quad \text{and}$$
$$c_{i=5-8} = (cos[(2i-9)\pi/4], sin[(2i-9)\pi/4])\sqrt{2}c, \quad (1)$$

where the lattice velocity $c = \Delta x/\Delta t$, Δx is the lattice size (uniform) and Δt is the duration of the time step. The rest particle corresponds to $i=0$ ($c_{i=0} = 0$) and others represent lattice vectors in the direction of the nearest neighbors. The complete velocity set of a D_2Q_9 model is listed in Table 1.

4.2. The Boltzmann Equation

The Boltzmann equation relates the time evolution and spatial variation of a collection of molecules to a collision operator that describes the interaction of the molecules,

$$\frac{\partial f}{\partial t} + c_i \cdot \nabla f = \frac{1}{\epsilon} J(f, f), \qquad (2)$$

where $\epsilon = Kn = l/L$ and $f \equiv f_i(x, c_i, t)$. The Knudsen number Kn is defined by the length scale ratio, where l is length scale in the mean free path and L is the characteristics macroscopic length. A continuous system satisfies small values of ϵ whereas the system is no longer continuous if ϵ is large. The form of the collision function, $J(f, f)$, can be derived under the variety of different assumptions. Conventionally, it is assumed that the density is relatively low so only binary collisions need be considered. In general, the molecules are also assumed to be uncorrelated before the collision takes place. Considering two particles, as originally suggested by Boltzmann, the collision integral can be written as

$$J(f, f) = \frac{1}{2m} \int_0^\pi d\phi \int_0^{2\pi} d\Phi \int dV \, B(\phi, V) \left(f_1' f_2' - f_1 f_2 \right), \qquad (3)$$

where m denotes mass of particle. The variables f' and f represent post- and pre-collision distribution functions where ϕ is the angle of an incoming particle and Φ is the angle in the plane perpendicular to the relative speed V (measured with respect to some arbitrary origin), as shown in Fig. 5.

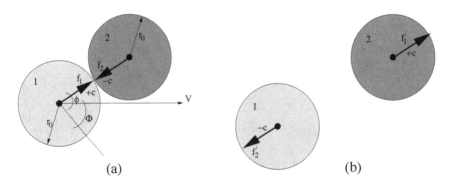

Fig. 5. Elastic collision between two hard-sphere molecules.

The quantity $B(\phi, V)$ is given by

$$B(\phi, V) d\phi \, d\Phi = \left(\frac{2\,k_b}{m}\right)^{-\frac{2}{s-1}} V^{\frac{(s-5)}{(s-1)}} \beta d\beta \, d\Phi, \qquad (4)$$

where k_b denotes the Boltzmann constant, $\beta = (mV^2/4\,k_b)^{1/s-1} 2r_0 sin\phi$ and r_0 denotes the radius of the sphere. We see that for the case $s = 5$, we have $B(\phi, V) = B(\phi)$. This leads to considerable simplifications in the theory of the Boltzmann equation. This result was first obtained by Maxwell, and molecules which are supposed to interact in this manner are referred to as Maxwell molecules. Therefore, for Maxwell molecules, the relaxation frequencies are independent of relative speed V.

In view of the mathematical difficulties posed in the solving the Maxwell-Boltzmann collision integral (3), investigators have suggested different levels of approximations of the collision processes. We shall spend some time looking into these approximations. However, our first point of investigations into applying the LB modeling approach will be the most popular one; the single time relaxation approximation (Bhatnagar et al. (1954)). Herein, we shall further assume that the molecules behave as hard spheres and collide elastically, as illustrated in Fig. 5.

4.3. Collision Model: Single-Time Relaxation

Computationally, the LB method belongs to a class of the pseudo-compressible solvers of the Navier-Stokes equations and can be classified as a Lagrangian, local equilibrium, finite-hyperbolicity approximation. The LBE method is originally derived from hyperbolic equations. The numerical formulation, as presented in the literature, often separates the equations into dissipative and non-dissipative parts when solution of the equation is sought. In the following and in the test cases presented herein, we consider a model in which the collision assumptions are simplified to a single-time relaxation form. The idea was first published by Bhatnagar et al. (1954). This is the simplest form of the LBE, and as mentioned, is referred to as the Bhatnagar-Gross-Krook (BGK) approximation. Since we focus on the Lattice BGK approach, we shall refer to this as the LBGK model. For details on work done on BGK modeling, the reader is referred to the research of Ghidaoui et al. (2006) who modeled shallow water flows. We should also mention the recent work of Liang et al. (2007) who describe surface water problems based on a BGK numerical model without need for any operator splitting.

The LBGK model implements a single time-relaxation-approximation collision model. The Boltzmann equation with the BGK collision model is discretized in velocity space by introduction of a finite set of velocities, c_i, and associated distribution functions, $f_i(x, c_i, t)$, yielding

$$\frac{\partial f_i}{\partial t} + c_i \cdot \nabla f_i = \int [f_1' f_2' - f_1 f_2] \approx -\frac{f_i - f_i^{eq}}{\tau}, \qquad (5)$$

where the first term represents the effect of the local change of the fluid motion in time and the second term describes the convection, that is, a linear advection operator. The first term on the right-hand side is the non-equilibrium distribution function which describes the effect of collisions (Succi (2001)). Discretizing equation (5) in space and time yields the following form, commonly known as the Lattice Boltzmann Equation (LBE):

$$f_i(\vec{x} + \Delta \vec{x}, t + \Delta t) = f_i(\vec{x}, t) - \frac{1}{\tau}\Big(f_i(\vec{x}, t) - f_i^{eq}(\vec{x}, t)\Big), \qquad (6)$$

where the velocity space is described through the finite set of velocities (1) and associated distribution functions (f_i) and equilibrium distribution functions (f_i^{eq}). The factor $1/\tau$ implements the kinematic viscosity,

$$\nu = \frac{c^2}{3}\left(\tau - \frac{\Delta t}{2}\right) \quad \text{where} \quad c = \frac{\Delta x}{\Delta t}, \qquad (7)$$

and is the inverse of the relaxation time, i.e. the collision frequency of the molecules or in other words the relaxation time represents the mean free time between collisions. The standard LBGK scheme of (5) is conditionally stable for $\tau > \Delta t/2$ and must be satisfied to ensure numerical stability. The definition of the the relaxation time also allows for the possibility of discretization error in kinetic viscosity. A useful discussion on viscosities in LBE can be found in the article by Dellar (2001). We should also recall that LBE actually are compressible but when the Mach number $Ma = u/c_s$ is small, then density (and temperature) fluctuations are $\mathbb{O}(Ma^2)$ and the flow is approximately isothermal and incompressible. The dynamic viscosity is independent of density and a function of temperature only. One should also bear in mind always to show that density fluctuations are minimized in LBE simulation tests. Fluid density time series, at carefully selected locations in the flow domain, can simply be used as a measure of demonstrating that numerical error has been minimized in an incompressible flow simulations, as it is constant for our ocean and coastal engineering applications of interest, that is, herein with focus on non-overturning waves. We should mention that recent discussions have been undertaken on the importance of compressibility in water waves during runup and impact on sea walls. Presentations on this matter was given by Prof. Peregrine, University of Bristol (U.K.), and Prof. Dalrymple, Johns Hopkins (USA), at the *ICCE* 2006 held in San Diego, USA. We are not entering this discussion herein but it should be mentioned that an LB model may be a useful contributing candidate on this matter. For now the reader is referred to the discussion by Dalrymple and Rogers (2006) who, in a series of publications, have demonstrated usage of the smoothed particle hydrodynamics approach to predict water wave behavior.

The equilibrium distribution function f_i^{eq} represents the invariant function under collision (no gradients are involved) and is dependent on the microscopic velocity of the molecule (c_i) and the macroscopic velocity (u). In classical kinetic theory the equilibrium distribution function f_{eq} is described by the Maxwellian equilibrium (collisions do not contribute)

$$f^{(0)} = \frac{\rho}{(2\pi\theta)^{D/2}} \exp\left[\frac{(c-\bar{u})^2}{2\theta}\right], \qquad (8)$$

where $c \equiv \Delta x/\Delta t = \sqrt{3\theta}$; or $c_s^2 = \theta = c^2/3$. Dimension of space is D, ρ, $\bar{u}$ and $\theta = k_B T/m$ are the macroscopic density of mass, the velocity and the normalized temperature where T is the absolute temperature, k_B is the Boltzmann constant and m is the particle mass. The Maxwell velocity distribution function (8), published in 1859, is considered to be a key result in developing a connection between microscopic and macroscopic physical processes. Note that Maxwell's distribution law (8) has identical form to a Gaussian distribution with average velocity u and variance c_s. In the LBM framework c_s is referred to as the speed of sound. However, the physical interpretation of c_s^2, should this be attempted in our water wave calculations, is only a measure of "the width of the bell" of the equilibrium distribution function. The choice of $c_s^2 = c^2/3$ is rooted in recovery of the Navier-Stokes equation. It is a free parameter in the LB schemes and can be tuned to a particular application of interest. We note that an expansion of $f^{(0)}$ and f up to $\mathbb{O}(u^2)$ is required to derive the Navier-Stokes equation,

$$f^{eq} = \frac{\rho}{(2\pi\theta)^{D/2}} \exp\left[-\frac{c^2}{2\theta}\right] \left(1 + \frac{c \cdot u}{\theta} + \frac{(c \cdot u)^2}{2\theta^2} - \frac{u^2}{2\theta}\right) + \mathbb{O}(u^3).$$

The general form of the equilibrium distribution functions can be expressed as a power series in macroscopic velocity (Rothman and Zaleski (1997)),

$$f_i^{eq} = A + B\, c_i\, u_i + C\, c_i\, c_j\, u_i\, u_j + D\, u_i\, u_j \delta_{ij}, \qquad (9)$$

where the constants $(A-D)$ are found by imposing conservation constraints (11) and δ_{ij} is the Kronecker delta. There are some degrees of freedom in the expressions of the equilibrium distribution but the moments must represent conservation of mass and momentum and static/dynamic pressures. It should also be noted that f_i is by definition an averaged, smooth, quantity. This stems from the original development of the nonlinear LBE in which noise is erased in the microdynamics of the LGCA scheme. The standard equilibrium distributions (without a free-surface) for the D_2Q_9 model are defined as

$$f_0^{eq} = \frac{4}{9}\left[\rho - \frac{3}{2}u \cdot u\right],$$

$$f_{i=1-4}^{eq} = \frac{1}{9}\left[\rho + 3c_i \cdot u + \frac{9}{2}(c_i \cdot u)^2 - \frac{3}{2}u \cdot u\right], \quad (10)$$

$$f_{i=5-8}^{eq} = \frac{1}{36}\left[\rho + 3c_i \cdot u + \frac{9}{2}(c_i \cdot u)^2 - \frac{3}{2}u \cdot u\right],$$

where ρ is the fluid density. The equilibrium distributions have important conservation constraints, yielding zeroth to second-order moments. For second-order accurate solutions, the conservation constraints have to be satisfied at each lattice site for each cartesian component. The moments represent conservation of mass and momentum and static/dynamic pressures, as follows:

$$\sum_{i=0}^{8} f_i^{eq} = \rho \,;$$

$$\sum_{i=0}^{8} c_i f_i^{eq} = \rho u \,; \quad (11)$$

$$\sum_{i=0}^{8} c_i c_j f_i^{eq} = \frac{1}{2}g\rho^2 \delta_{ij} + \rho u u \,,$$

where g is the acceleration of gravity. The density in (11) becomes the total water depth h in the shallow water applications, as shown later. Satisfying the equilibrium distribution functions is essential in obtaining valid physics. Note that the form of the functions would be dependent on the application of interest. For example, the deep water formulation would have different functions compared to the shallow water equation counter-part. These functions would typically have to be derived through Taylor expansions and Chapman-Enskog analysis from which it can be shown that the Navier-Stokes equation can be derived from the Boltzmann equation. This kind of analysis will be described in the following §4.4.

4.4. LB and Navier-Stokes Flows

There are mainly two ways to truncate an infinite series of transport equations. As mentioned, one technique is based on the Chapman-Enskog method and another is introduced by Grad (1949). Herein we highlight the former method. Chapman (1888–1970) in England and Enskog (1884–1947) in Sweden described independently the general solution to the Boltzmann equation in a series of papers published 1911–21, e.g. Chapman (1916, 1918); Enskog (1921). The primary use is the calculation of transport equations. As we know these quantities are unknown in the macroscopic theories. The Chapman-Enskog procedure considers an expansion of the equations for the moments of f_i and assumes that the time dependence of f_i occurs through mass density, velocity and temperature. The Chapman-Enskog expansion corresponds to a multi-scale expansion to first-order in space and second-order in time and assumes that the diffusion time scale is much slower than the convection time scale.

Below we shall briefly demonstrate that the Boltzmann equation (2) satisfies the Navier-Stokes equation using the second-order Chapman-Enskog expansion technique. We assume the moments of the equilibrium distribution is equal to the Maxwell-Boltzmann distribution. The analysis is performed for the Lattice BGK scheme. The Chapman-Enskog expansion parameter is the Knudsen number (ϵ). Our definition of the coordinate system and the subscript indices follow the velocity set of the LB cube, as shown in Fig. 4(b).

To perform the Chapman-Enskog expansion, a Taylor expansion of the discretized Boltzmann equation (6) is first undertaken,

$$f_i(\vec{x} + \Delta\vec{x}, t + \Delta t) \approx \left(1 + D + \frac{1}{2}D^2 + \frac{1}{6}D^3 + ...\right) f_i(\vec{x}, t). \tag{12}$$

The differential operator $D = \Delta t (c_{i,\alpha} \partial_\alpha + \partial_t)$ where the first term denotes the spatial derivatives and the second term is the time derivative. Inserting the expression of D into (12) up to second-order, the Taylor expansion is

$$f_i(\vec{x} + \Delta\vec{x}, t + \Delta t) - f_i(\vec{x}, t)$$
$$\approx \left[\partial_t + c_{i,\alpha}\partial_\alpha + \frac{1}{2}c_{i,\alpha}\partial_\alpha(c_{i,\beta}\partial_\beta + \partial_t) + \frac{1}{2}\partial_t(c_{i,\alpha}\partial_\alpha + \partial_t)\right] f_i(\vec{x}, t)\, \Delta t. \tag{13}$$

For simplicity reasons, the notation $(\partial_{1x})_\alpha = \partial_{1\alpha}$. This also applies to the y- and z-directions.

Expanding the distribution functions, the time and space derivatives in terms of ϵ about equilibrium gives

$$f_i = f_i^{(0)} + \epsilon f_i^{(1)} + \epsilon^2 f_i^{(2)} + \ldots ;$$
$$\Delta t\, \partial_t = \epsilon \partial_{1t} + \epsilon^2 \partial_{2t} + \ldots ; \qquad (14)$$
$$\Delta t\, \partial_\alpha = \epsilon \partial_{1\alpha}.$$

Note that weak deviations from equilibrium is assumed ($\epsilon << 1$; $\epsilon \approx \Delta t$). The zeroth-order approximation, $f_i^{(0)}$, is taken to be the equilibrium distribution, f_i^{eq}. We note that convection and diffusion processes are assumed to operate at different time scales; t_1 denotes convection and is the fast time scale whereas the slower time scale t_2 relates to diffusion.

We can perform the Chapman-Enskog expansion by substituting (14) into (6) and (13), resulting in the following expression,

$$\left[(\epsilon\partial_{1t} + \epsilon^2\partial_{2t}) + c_{i,\alpha}\epsilon\partial_{1\alpha} + \frac{1}{2}c_{i,\alpha}\epsilon\partial_{1\alpha}(c_{i,\beta}\epsilon\partial_{1\beta} + (\epsilon\partial_{1t} + \epsilon^2\partial_{2t}))\right.$$
$$\left. + \frac{1}{2}(\epsilon\partial_{1t} + \epsilon^2\partial_{2t})(c_{i,\alpha}\epsilon\partial_{1\alpha} + (\epsilon\partial_{1t} + \epsilon^2\partial_{2t}))\right] \times \left(f_i^{(0)} + \epsilon f_i^{(1)} + \epsilon^2 f_i^{(2)}\right)$$
$$\approx -\frac{1}{\tau}\left(f_i^{(0)} + \epsilon f_i^{(1)} + \epsilon^2 f_i^{(2)} - f_i^{eq}\right). \qquad (15)$$

The conservation of mass and momentum requires $\sum_i f_i^{(a)} = 0$ and $\sum_i f_i^{(a)} c_{i,\alpha} = 0$ where $a = 1, 2$. To first-order in ϵ, (15) reduces to

$$\partial_{1t} f_i^{(0)} + c_{i,\alpha}\, \partial_{1\alpha} f_i^{(0)} \approx -\frac{1}{\tau} f_i^{(1)}. \qquad (16)$$

Summing equation (16) using (11) gives

$$\partial_{1t}\rho + \partial_{1\alpha}\rho u_\alpha = 0. \qquad (17)$$

Multiply equation (16) by $c_{i,\beta}$ to get

$$c_{i,\beta}\, \partial_{1t} f_i^{(0)} + c_{i,\beta}\, c_{i,\alpha}\, \partial_{1\alpha} f_i^{(0)} \approx -\frac{1}{\tau} f_i^{(1)} c_{i,\beta}. \qquad (18)$$

Summing (18) and using (11) gives

$$\partial_{1t}\rho u_\beta + \partial_{1\alpha}\rho u_\alpha u_\beta \approx -\partial_{1\alpha} \frac{1-d_0}{D_n} \rho c^2 \delta_{\alpha\beta}, \qquad (19)$$

where d_0 is a weighting coefficient related to the rest particle. It is defined as $d_0 = \rho_0/(q+1)$ where q is the total number of states on a lattice site and ρ_0 is the average particle density, as introduced by Chen et al. (1992) (for example, D_1Q_3: $d_0 = 1/3\rho$; D_2Q_9: $d_0 = 4/9\rho$; D_3Q_{19}: $d_0 = 1/3\rho$). D_n denotes the dimensional space. The term $\delta_{\alpha\beta}$ is the Kronecker delta function where $\delta_{\alpha\beta} = 1$ if $\alpha = \beta$ or $\delta_{\alpha\beta} = 0$ if $\alpha \neq \beta$. Expansion up to second-order in ϵ, we get

$$\partial_{2t} f_i^{(0)} + \partial_{1t} f_i^{(1)} + c_{i,\alpha} \partial_{1\alpha} f_i^{(1)} + \frac{1}{2}\partial_{1t}\left[\partial_{1t} f_i^{(0)} + c_{i,\alpha} \partial_{1\alpha} f_i^{(0)}\right]$$
$$+\frac{1}{2}\partial_{1\alpha}\left[c_{i,\alpha} \partial_{1t} f_i^{(0)} + c_{i,\beta} c_{i,\alpha} \partial_{1\beta} f_i^{(0)}\right] \approx -\frac{1}{\tau}f_i^{(2)}. \qquad (20)$$

Summing (20) over i, it can be observed that several terms are zero due to the mass and momentum conservation and due to (17) and (19). This leaves

$$\partial_{2t}\rho = 0. \qquad (21)$$

Multiplying (20) by $c_{i\gamma}$ and summing over i gives

$$\sum_i \left(c_{i,\gamma} \partial_{2t} f_i^{(0)} + c_{i,\gamma} \partial_{1t} f_i^{(1)} + c_{i,\gamma} c_{i,\alpha} \partial_{1\alpha} f_i^{(1)} \right.$$
$$+\frac{1}{2}\partial_{1t}\left[c_{i,\gamma} \partial_{1t} f_i^{(0)} + c_{i,\gamma} c_{i,\alpha} \partial_{1\alpha} f_i^{(0)}\right] \qquad (22)$$
$$\left. +\frac{1}{2}\partial_{1\alpha}\left[c_{i,\alpha} c_{i,\gamma} \partial_{1t} f_i^{(0)} + c_{i,\gamma} c_{i,\beta} c_{i,\alpha} \partial_{1\beta} f_i^{(0)}\right] \approx -\frac{1}{\tau}f_i^{(2)} \right).$$

Several terms are can be found to be zero, yielding

$$\partial_{1\alpha} \sum_i c_{i,\alpha} c_{i,\gamma} f_i^{(1)}$$
$$= -\tau\left(\frac{(1-d_0)c^2}{D_n}\partial_{1t}\partial_{1\alpha}\rho\delta_{\alpha\beta} + \partial_{1\alpha}\partial_{1\beta}\frac{\rho c^2}{D_n+2}[u_\alpha\delta_{\beta\gamma} + u_\beta\delta_{\alpha\gamma} + u_\gamma\delta_{\alpha\beta}]\right). \qquad (23)$$

One can further show that

$$\partial_{2t}\rho u_\gamma = \nu\partial_{1\alpha}\partial_{1\alpha}\rho u_\gamma + \partial_{1\gamma}(\Upsilon\partial_{1\alpha}\rho u_\alpha), \qquad (24)$$

where $\nu = c^2(\tau - 1/2)/(D_n + 2)$ denotes the kinematic viscosity and $\Upsilon = (\tau - 1/2)[2c^2/(D_n+2) - c^2(1-d_n)/D_n]$ is the bulk viscosity. Now the continuity equation reads

$$\partial_t \rho + \partial_\alpha \rho u_\alpha = 0, \qquad (25)$$

and the Navier-Stokes equation is,

$$\partial_t \rho u_\alpha + \partial_\beta \rho u_\beta u_\alpha = \partial_\beta\left[\frac{\rho(1-d_0)}{D_n}c^2\delta_{\alpha\beta}\right] + \nu\partial_\beta\partial_\beta\rho u_\alpha + \partial_\alpha \Upsilon \partial_\beta \rho u_\beta. \qquad (26)$$

We should also emphasize that the Chapman-Enskog technique applied to show that LBE satisfies the Navier-Stokes equation is widely published. The reader is invited to extract further details on this topic in the literature: in books, e.g. Chapman and Cowling (1958); Vincenti and Kruger, Jr. (1965); Gombosi (1994); Struchtrup (2005); in papers, e.g. Koelman (1991); Chen et al. (1992); Qian et al. (1992).

4.5. Boundary Treatments

In the LB method, the boundary conditions are described through distribution functions f_i. There exists a variety of schemes used for implementation of wall/bed boundary treatments. The most popular one is "the bounce-back scheme", e.g. Zhou (2004) which includes cells outside the computational domain (ghost cells). In this scheme, particles bounce against the wall and the particle distribution are scattered back to the node it came from. An example of a free-slip boundary is shown in Fig. 6. Suggestions to improve the bounce-back scheme are described by Ginzburg and

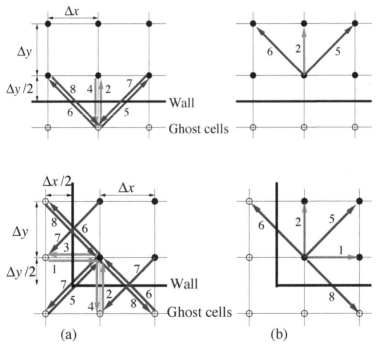

Fig. 6. Free slip boundary conditions at wall and corner (a) before and (b) after streaming.

D'Humières (2003); Noble et al. (1995). Recent investigations on boundary conditions treatment are discussed by Pan et al. (2006).

It is notable that the present depth integrated LBGK model does not include the conventional dynamic and kinematic boundary conditions at the free-surface. Instead, the free-surface dynamics are accounted for through the non-equilibrium particle distribution function ($f^{neq} = f_i - f_i^{eq}$). No additional algorithm or surface boundary conditions are necessary to be prescribed in the present numerical model. We should stress that the present model can only expect to work well for a continuous surface in shallow water. Discontinuous surfaces and/or other water depths would involve entirely new LB formulations. In particular, new sets of equilibrium distribution functions would be needed. A numerical Boltzmann formulation in arbitrary water depth could include a free-surface treatment as known from traditional solvers, e.g. a volume of fluid algorithm, a level set, etc. to account for breaking. Therefore a host of models could potentially be developed to bridge the gap between micro and macroscopic models and thus potentially advance the way we currently model free-surface flows. A further strength is that the standard LBE scheme is known not to introduce numerical viscosity.

Regarding open boundary and wet-dry interface treatments, please see the description related to the wave runup test case presented in Chapter 16 in this volume.

4.6. Standard LBE and Higher-Order Schemes

With reference to the list of LBM properties in the introduction section, we are now ready to make some further important observations. First we recall that Eq. (6) is referred to as the standard LBE. We observe that this is the results of replacing the time derivatives (5) by a first-order time difference and a first-order upwind discretization for the convective term. We observe that (6) represents a finite difference equation. Therefore, viewing the LBM as a finite-difference model for solving the discrete-velocity Boltzmann equations, it becomes necessary to address numerical instability and accuracy. It is important to note that the equation is first-order accurate in time and space. This contradicts the former derivation (§4.4) as it was shown that the LBE satisfies the Navier-Stokes equation in the near incompressible limit. Sterling and Chen (1996) mentioned that the LBE is second-order accurate both in time and space due to the Lagrangian nature of the spatial discretization of (6) embedded in the viscous term. Recently,

Junk *et al.* (2005) undertook a detailed analysis of the LBE. They related the LBE to the finite discrete-velocity model of the Boltzmann equation with diffusive scaling and obtained the incompressible Navier-Stokes equations as opposed to the compressible Navier-Stokes equations obtained by the Chapman-Enskog analysis with convective scaling. Junk *et al.* proved that the LBE is second-order accurate in space. For the incompressible Navier-Stokes, they also proved that the LBE is first-order accurate in time asymptotically. Gou *et al.* (in press) also recently contributed to this concern. Through unsteady flow predictions, Guo *et al.* observed that LBE is second-order accurate in time.

Further, we note that the standard bounce-back condition, the most applied boundary condition, is first-order in numerical accuracy at the boundaries. This, of course, degrades the LB method. As mentioned in the review, several investigators have contributed to second-order accurate boundary conditions. New contributions to the LBM literature are for this reason (and others) still on-going research.

In this chapter, we present results based on the standard LBE. The wave runup test case, presented in Chapter 16 in this volume, has undergone further tests including high-order finite-discretization schemes, thus improving the standard LBE. We should mention that it did not originally occur to investigators that the LBE could be expanded this way (like traditional CFD methods). It is still not common to consider a particular discretization for the discrete Boltzmann equation. However, some progress on high-order LB discretization schemes have been reported. Cao *et al.* (1997) were the first investigators to view the LBM as a special finite difference (FD) discretization. Further elaborations on this topic have been undertaken since, e.g. Kandhai *et al.* (2000); Junk (2001); Seta and Takahashi (2002); Sofonea and Sekerka (2003); Van der Sman (2006). Some Finite Volume (FV) LB models have also been developed, e.g. Ghidaoui *et al.* (2006); Que and Xu (2006). As far was we are aware of, there has been no comparison undertaken for FDLB and FVLB model, so the difference in solutions accuracy is not clear to us. Furthermore, we should also mention that various discretization schemes for gas dynamics computations are described in details in the book by Laney (1998).

4.7. *Gas Dynamics Analogy*

Since it is not obvious why a numerical free-surface model with roots in gas dynamics theory would work for water waves, we have gathered

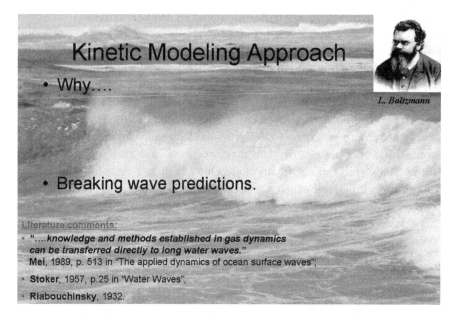

Fig. 7. Kinetic modeling approach to water wave predictions with roots in the gas dynamics theory of Boltzmann. Comments and equations derivations on this idea can be found in the listed summary of references. (The photograph is a wave runup snapshot from the Northshore of Oahu, Hawai'i.)

some interesting comments and observations made by other investigators, as summarized in Fig. 7. It is believed that Dimitri Riabouchinsky (1882–1962), a Muscovite fluid dynamicist, was the first investigator to publish a note on recognizing the similarity between the equations for shallow water theory and the fundamental differential equations of gas dynamics (Riabouchinsky (1932)). Stoker (1957) elaborates on the French notes and has a couple of pages in his book (pp. 25–26) on this subject. The book of Mei (1992) also has some related notes (pp. 512–513) on this topic and one can find a quote: "Knowledge and methods established in gas dynamics can be transferred directly to long water waves."

Although the LB results herein simulate long waves, we should say that the reason why we have proposed the LBM originates from independent and different ideas. It is our goal and intent to apply the method to breaking waves in any water depth. It was nevertheless very encouraging to find these references which are relevant to the present contribution.

5. LBGK Modeling with Application to Water Waves

We shall in the remaining part of the chapter focus on applying and testing a Lattice Boltzmann modeling approach to free-surface water waves. We assume that the collisions of particles are represented through the single time relaxation scheme (LBGK approximation). It should be noted that there is no deeper reason for starting the Boltzmann implementation this way other than this is the first logical modeling development step. However, this is not necessarily the most accurate and stable approach due to the BGK collision approximation. Our LB model contains equilibrium distribution functions which are based on particle motions in shallow water and therefore our test cases are limited to shallow water depths. The discretized equations are based on viscous free-surface waves. Therefore, flows with horizontal velocity dominating free-surface waves is applicable and is our focus only. Furthermore, the present LBGK model explores the free-surface behavior of non-overturning water waves on uniform grids. For validation purposes, we first investigate the dynamics of viscous liquid and water waves in tanks with flat and slope beds; in 1-D and 2-D, respectively. Then, we test the tsunami runup on sloped beaches (workshop benchmark test 1). The latter test case is presented in Chapter 16 in this volume.

We consider the flow of water with a free-surface under gravity in a 3-D domain where $x - y$ denotes a horizontal plane while z defines the vertical direction. The free-surface elevation coincides with the z-axis. Assuming that the vertical component is negligible in the field of interest, the fluid motion can be described by the shallow water equations with rotating flows,

$$\frac{\partial h}{\partial t} + \frac{\partial (h\, u_x)}{\partial x} + \frac{\partial (h\, u_y)}{\partial y} = 0\,;$$

$$\frac{\partial (h\, u_x)}{\partial t} + \frac{\partial (h\, u_x^2)}{\partial x} + \frac{\partial (h\, u_x u_y)}{\partial y}$$

$$= -g\frac{\partial}{\partial x}\left(\frac{h^2}{2}\right) + \frac{\partial}{\partial x}\left(h\nu \frac{\partial u_x}{\partial x}\right) + \frac{\partial}{\partial y}\left(h\nu \frac{\partial u_x}{\partial y}\right) + F_x\,; \quad (27)$$

$$\frac{\partial (h\, u_y)}{\partial t} + \frac{\partial (h\, u_y^2)}{\partial y} + \frac{\partial (h\, u_x u_y)}{\partial x}$$

$$= -g\frac{\partial}{\partial y}\left(\frac{h^2}{2}\right) + \frac{\partial}{\partial x}\left(h\nu \frac{\partial u_y}{\partial x}\right) + \frac{\partial}{\partial y}\left(h\nu \frac{\partial u_y}{\partial y}\right) + F_y\,,$$

where $h = h_0 + \zeta$, h_0 is the still water depth (or initial water depth) and ζ denotes the free-surface elevation measured vertically above still water level, t is time, $g = 9.81$ m/s^2 denotes the acceleration of gravity, $\rho = 1000$ kg/m^3 is the water density, $\nu = 1 \times 10^{-6}$ m^2/s is the kinematic viscosity, u_x, u_y are the horizontal depth-averaged velocity components, and F denotes force terms.

Equivalent to (27), the proposed single phase LBGK formulation for shallow liquid rotating flows can be written as

$$\frac{\partial f_i}{\partial t} + c_i \frac{\partial f_i}{\partial x} = -\frac{(f_i - f_i^{eq})}{\tau} + \frac{1}{N_i c^2} c_i F_i, \quad \text{where} \quad N_i = \frac{1}{c^2} \sum_{i=0}^{2} c_i c_i, \tag{28}$$

with N_i as a constant which depends on the lattice geometry and is defined by $N_i = 2$ and $N_i = 6$ for the D_1Q_3 and D_2Q_9 velocity set, respectively. F denotes force terms. Herein, it is assumed that, at any time, the LB fluid is characterized by the populations of three discrete velocities (D_1Q_3 model) in 1-D. The nine discrete microscopic velocity model (D_2Q_9 model) represents the physics of waves in a 2-D horizontal plane ($x - y$) based on the depth-average equations (27), as shown in Fig. 2.

Equation (28) can be viewed as a special finite-difference discretization of the single time relaxation approximation of the Boltzmann equation for discrete velocities. One approach to solving the discrete Boltzmann equations is to use a first-order Euler time difference scheme and a first-order upwind space discretization for the convection term in a uniform lattice spacing Δx. One can then obtain following algebraic relation,

$$f_i(\vec{x} + \Delta \vec{x}, t + \Delta t) - f_i(\vec{x}, t) = -\frac{\Delta t}{\tau} \Big(f_i(\vec{x}, t) - f_i^{eq}(\vec{x}, t) \Big) + \frac{\Delta t}{N_i c^2} c_i F_i. \tag{29}$$

This equation is commonly referred to as the Lattice-BGK equation or Lattice-Boltzmann Equation (LBE). As mentioned, it can be shown that the LBE is second-order accurate in space and first-order accurate in time (Junk et al. (2005)). See also further discussion of this in §4.6.

The LBGK-based equilibrium distribution functions in shallow water for a D_2Q_9 model are (after Salmon (1999a)),

$$f^{eq}_{i=0} = h_0 + \zeta - \frac{5g(h_0+\zeta)^2 \Delta t^2}{6\Delta x^2} - \frac{2(h_0+\zeta)\Delta t^2}{3\Delta x^2}(u_i u_i) ,$$

$$f^{eq}_{i=1-4} = \frac{g(h_0+\zeta)^2 \Delta t^2}{6\Delta x^2} + \frac{(h_0+\zeta)\Delta t^2}{3\Delta x^2}(c_i u_i)$$

$$+ \frac{(h_0+\zeta)\Delta t^4}{2\Delta x^4}(c_i c_j u_i u_j) - \frac{(h_0+\zeta)\Delta t^2}{6\Delta x^2}(u_i u_i) , \quad (30)$$

$$f^{eq}_{i=5-8} = \frac{g(h_0+\zeta)^2 \Delta t^2}{24\Delta x^2} + \frac{(h_0+\zeta)\Delta t^2}{12\Delta x^2}(c_i u_i)$$

$$+ \frac{(h_0+\zeta)\Delta t^4}{8\Delta x^4}(c_i c_j u_i u_j) - \frac{(h_0+\zeta)\Delta t^2}{24\Delta x^2}(u_i u_i) ,$$

where the subscript i denotes the propagating direction of the discrete velocities, as described in Eqs. (1). Important to water waves simulations are the zeroth to second moments which represent conservation of mass and momentum and static/dynamic pressures, as follows:

$$\sum_i f^{eq}_i = h(x,t) ; \quad \sum_i c_i f^{eq}_i = h(x,t) u_i ;$$

$$\sum_i c_i c_j f^{eq}_i = \frac{1}{2} g h(x,t)^2 \delta_{ij} + h(x,t) u_i u_j . \quad (31)$$

An elastic-collision scheme assuming no flow through slip walls and bed have been prescribed using the well-known bounce-back rules (e.g. Zhou (2004)).

The present mesoscopic numerical solutions are solved on uniform lattices. Initially we typically prescribe the non-physical values of the velocity $u = 0$ and the distribution functions $f_i = f^{eq}_i$. In the test cases of the present chapter, we have used a splitting operator scheme, as outlined in the solution procedure of Fig. 8. So the standard LBE scheme herein separates into advection and diffusive parts before computing f_i. Finally, the microscopic properties are transformed into macroscopic variables of the free-surface (ζ) and the depth averaged velocities (u_x, u_y) are calculated as the first and second moments of the distribution function,

$$\zeta = \sum_i^8 f_i - h_0 ; \quad u_x = \frac{1}{(h_0+\zeta)} \sum_i^8 c_{ix} f_i \quad \text{and} \quad u_y = \frac{1}{(h_0+\zeta)} \sum_i^8 c_{iy} f_i . \quad (32)$$

The procedure is repeated at each time step Δt and the surface elevation and velocity is updated from (32). The hydrodynamic moments of the

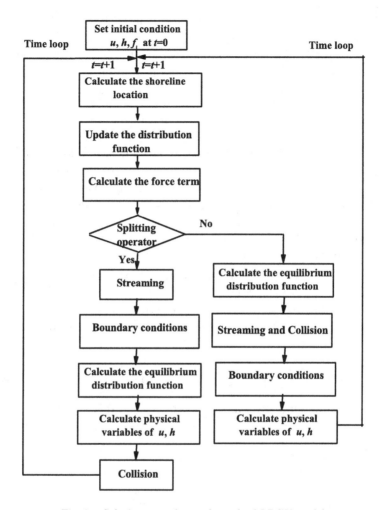

Fig. 8. Solution procedures of standard LBGK model.

equilibrium distribution functions (30) are satisfied at every time step. Further, it should be noted that it is assumed that the Mach number $Ma = u/c_s \ll 1$, where c_s is the speed of sound. The Peclet number $Pe = u\,\Delta x/\nu < 2$ and the Courant number $Cr = u\,\Delta t/\Delta x < 1$ are also obeyed ($u = u_x$ or $u = u_y$). It should be noted that the Pe and Cr constraints are borrowed from continuum level models and thus may not necessarily apply to LB models. Establishment of LB constraints requires more validation work than presented herein.

Finally, we should emphasize that the standard LBGK solutions of (28) are second-order accurate in space and first-order accurate in time (see discussion in §4.6). Further, it should be noted that the LBGK schemes are highly suited for vectorization and parallelization. However, the test cases herein are generated using a single CPU of about 3 GHz.

6. Case Studies

We have undertaken various test cases to examine the performance of the LBGK free-surface solver. These are preliminary test cases which do not involve the benchmark test cases of the workshop. We do this because the LB modeling approach for water waves has not undergone any thorough validation other than the 1-D LBGK studies in tanks undertaken by Frandsen (2006a). We should note that these test cases proved very helpful in the model development of the tsunami-generated wave runup modeling problem. An overview of the various numerical experiments undertaken so far are illustrated in Figs. 9–11.

Prior to the free-surface simulation test case initiation, we tested the LBGK solver performance without a free-surface, as described in

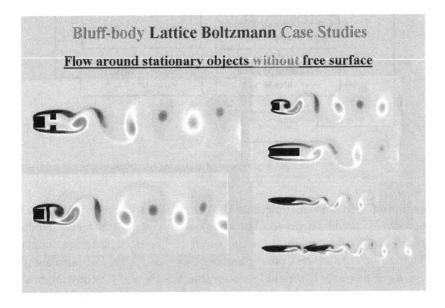

Fig. 9. LBGK bluff-body Navier-Stokes flows without a free-surface.

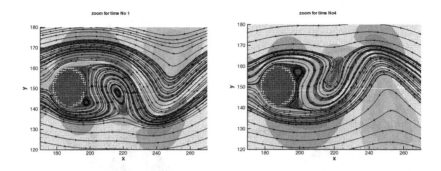

Fig. 10. LBGK flow past circular cylinder. Velocity snapshots ($Re = 200$).

Frandsen (2006b). This particular article also showed some initial attempts on predictions of free-surface bluff-body interactions (e.g. Fig. 2). These studies included single and multiple bluff-bodies studies at $Re \leq 250$ (Fig. 9). In general, we found that the LB model provided an easy means of handling multiple bluff-bodies with sharp edges. We are currently modeling flow around circular cylinders (Fig. 10). The curved boundary is modeled through the second-order accurate boundary condition proposed by Filippova and Hänel (1997). Good agreement with other investigators have been achieved for $Re \leq 250$. We have also implemented a curvilinear formulation, as developed by He and Doolen (1997). This approach seems to work equally well for flow around circular cylinder.

The free-surface test cases, as illustrated in Fig. 11, have so far included (1) weak bore simulations; (2) simulations of wave interactions with surface-piercing bodies; (3) dam-break predictions; (4) force term assessment (e.g. beach slope); and (5) wave runup. We have done extensive studies of bore predictions (without bluff-bodies) and wave runup.

We shall in the following, present some of the bore studies and dam-break predictions. The wave runup studies follow the description of the benchmark test 1 set-up problem, which is reported in Chapter 16 in this volume.

6.1. Case Study: Viscous Based Bores in Fixed Tanks

Viscous free sloshing in fixed rectangular tanks is chosen as a first benchmark validation test of the LBGK solver. The numerical model is based on the NLSW in rotational flow (27). Our results are based on a D_2Q_9

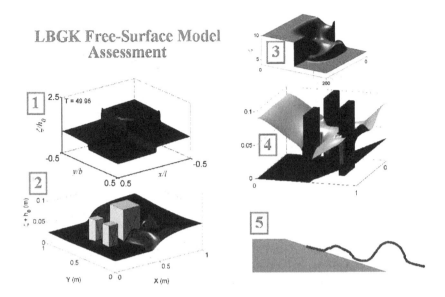

Fig. 11. LBGK free-surface flows tests in shallow water.

model in the horizontal plane $(x\text{-}y)$. The bounce-back scheme for slip walls have been prescribed. We should emphasize that the solutions in this first test series are based on highly viscous liquid. This is not only motivated by our comparison source (Wu *et al* (2001)) but also because we were at first not certain if we could simulate free-surface behavior at all, including both highly viscous liquid and water. Therefore, we began our studies with highly viscous liquid.

The numerical predictions are compared with the linearized Navier-Stokes solution derived by Wu *et al.* (2001) who showed that the wave elevation at an arbitrary depth is given by

$$\zeta = \zeta_0 + \frac{\zeta_0\, g\, k_2\, tanh(k_2\, h_0)}{2\pi\, \nu\, i} \int_\Gamma \frac{tanh(\alpha_2\, h_0)}{C_2}\, e^{\lambda t}\, d\lambda\,, \qquad (33)$$

where $k_2 = 2\pi/b$ is the wave number for n_2 and $\alpha_{n=2} = \sqrt{k_n^2 + \lambda/\nu}$, where λ is a complex number and b denotes the width of the tank. The denominator in (33) is complex and is presented in detail in the article of Wu *et al.* (an analytical solution exists for $\lambda = 0$ only). The deep water limit of (33) is explored by Wu *et al.* (2001) for $Re = 2, 20, 200$, with $Re = h_0\sqrt{g\,h_0}/\nu$, in a rectangular tank with an aspect ratio $h_0/b = 0.5$. In this paper, it is the shallow liquid limit of (33) which is used in the comparison studies. Assuming that the liquid depth is sufficiently small such that $tanh(k_n\, h_0) \to k_n\, h_0$ and $tanh(\alpha_n\, h_0) \to \alpha_n\, h_0$, it can be shown that the denominator simplifies to

$$C_2 = -\frac{\lambda\, \alpha_2\, h_0}{\nu}\left(\lambda^2 + 4 k_2^2 \nu \lambda + g\, k_2^2\, h_0\right), \qquad (34)$$

and (33) simplifies to

$$\zeta = \zeta_0 + \frac{\zeta_0\, g\, k_2^2\, h_0}{2\pi\, \nu\, i} \times \int_\Gamma \frac{\nu\, e^{\lambda t}}{-\lambda^3 - 16\pi^2\, \nu \lambda^2 + g\, 4\pi^2\, h_0\, \lambda}\, e^{\lambda t}\, d\lambda. \qquad (35)$$

In this form, the denominator has three poles in the complex plane which allows the integral to be calculated analytically. The integration path Γ is the path for calculating inverse Laplace transforms. Letting $\omega_n^2 = g\, k_n^2\, h_0$ where ω_n represents the sloshing frequency in shallow liquid, (35) reduces to

$$\zeta = \zeta_0 + \frac{\zeta_0\, \omega_n^2}{k_n^2\, \nu}\, e^{-\nu\, k_n^2\, t}\, f(\nu\, k_n^2\, t), \qquad (36)$$

where $f(\nu\, k_n^2\, t)$ is the inverse Laplace transform which solution can be found in Poularikas (1996). The derivation leads to following solution for the viscous time evolution of the free-surface in shallow liquid,

$$\begin{aligned}\zeta = \frac{\zeta_0}{2\sqrt{4 - \omega_n^2/(k_n^2 \nu)^2}} &\Big[1 - k_n^2\, \nu \\
&- k_n^2\, \nu \left(2 + \sqrt{4 - \omega_n^2/(k_n^2 \nu)^2}\right) e^{\left(-2 - \sqrt{4 - \omega_n^2/(k_n^2 \nu)^2}\right)\nu\, k_n^2\, t} \\
&+ k_n^2\, \nu \left(2 - \sqrt{4 - \omega_n^2/(k_n^2 \nu)^2}\right) e^{\left(-2 + \sqrt{4 - \omega_n^2/(k_n^2 \nu)^2}\right)\nu\, k_n^2\, t}\Big].\end{aligned} \qquad (37)$$

Initially the velocity in the flow domain is zero, and the free-surface $\zeta_0(x, n) = a\, cos(k_n x)$, where a is the amplitude of the initial wave profile and x is the horizontal distance from the left wall. Nonlinear free-surface motions are investigated by varying the wave steepness, defined herein as $\epsilon = a\, \omega_n^2/g$ and is a measure of non-linearity at the free-surface in the fixed-tank studies. The results presented are for a tank of aspect ratio h_0/b = 0.05 and for Re = 20, 200 with corresponding $\nu = 1.75 \times 10^{-3}$ m^{2}s^{-1}, 1.75×10^{-4} m^{2}s^{-1}. The Re = 20 corresponds to glycerine type of liquid whereas Re = 200 is oil-based liquid (SAE 20W-20). The time histories of the free-sloshing motions are presented in non-dimensional form using the sloshing frequency ω_n and the initial amplitude a.

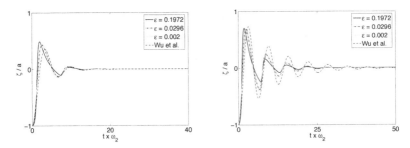

Fig. 12. Free-surface at wall at $Re = 20$ and $Re = 200$.

Figure 12 shows the time histories of the free-surface elevation for the second sloshing mode (n = 2) for the liquids with $Re = 20$ and 200, respectively. Each of these Re solutions are based on a grid resolution of 100 and 500 nodes, respectively. Solutions are presented for time steps Δt = 0.01 s, 0.002 s and Δx = 0.01 m, 0.002 m yielding $\tau = 1.025$ for lowest grid density/Re case and $\tau = 0.7625$ for highest grid density/Re case. The increase of nodes and associated reduced time step were necessary in order to maintain accuracy of the explicit solver for higher Re. The free-surface time histories at the left wall are shown for small to moderate waves ($\epsilon = 0.002, 0.197$). The $Re = 20$ result shows one oscillation cycle only. Hereafter, due to the high viscosity, the amplitude decays to zero. This is also true for all of the other wave steepness cases considered shown. The small initial wave $\epsilon = 0.002$ case agrees fairly well with the linear solution of Wu et al. The increasing wave steepness introduces a phase-shift with larger peaks and lower troughs. The higher $Re = 200$ test case generates larger wave steepness ($k_2 \times \zeta_{max} = 0.43$) with associated amplitude skewness and an increase in oscillation cycles, which eventually decays toward zero, representing the physics well. We should point out that the skewness of the amplitudes is the characteristic of a bore as opposed to non-skewed amplitudes of sloshing motion. Bores are known to form at shallow liquid depths for a relatively large forcing frequency. The next validation study is on water bores.

6.2. *Case Study: Water Based Bores in Moving Tanks*

Recently, the author was involved in setting-up small scale wave tank experiments to investigate sloshing motions. The test facility is shown in Fig. 13. The set of test series described below were highly motivated by having this experimental facility. The present model explores the free-surface behavior

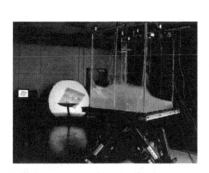 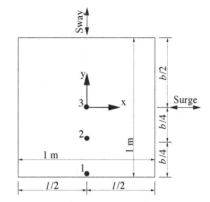

Fig. 13. Shake table with shallow water sloshing in a square tank with base dimensions of 1×1 m². Definition of excitation modes and plan view of tank with wave gauge locations (1–3). (The photograph is from the laboratory of Frandsen located at Louisiana State University, USA.)

of the LBGK model, as described above. We have tested the performance of the LBGK free-surface solver in tanks on flat beds ($F_x = 0; F_y \neq 0$ in Eq. 27). The forced horizontal acceleration of the tank in the y-direction (sway) is $Y'' = a_h \omega_h^2 \cos(\omega_h t)$ where a_h, ω_h are forcing amplitude and frequency, respectively ($F = h \times Y''$ in Eq. 28). The case study presented includes free-surface behavior with an initial water depth $h_0/b = 0.05$ where $b = 1$ m is the tank width. At this shallow water depth, it is expected that the free-surface has horizontal dominating velocity components, and thus is suitable for the present model formulation. Moreover, we emphasize that in shallow water depth, a traveling wave or a bore may be generated, as reported by, e.g. Wu et al. (1998) and Mei (1992), as opposed to sloshing motion occurring in moderate to deep water depths. Herein we discuss the free-surface behavior in relation to waves generated due to sway (and/or surge) base excitation in a square tank, as illustrated in the sketches of Figs. 13 and 14. Special attention is given to cases that generate bores. Although the bore formation is confined in tanks, the physics at the location of the bore is similar to bores observed on beaches, as researched by, e.g. Keller et al. (1960); Shen and Meyer (1963); Hibberd and Peregrine (1979); Yeh et al. (1989). Moreover, fully developed tank bores resemble the tongue of a tsunami wave. However, due to the non-overturning numerical formulation, our numerical study is limited to weak bores.

The effects on the free-surface when forcing the tank to move in moderate sway motion, are investigated. The forcing frequency is varied to

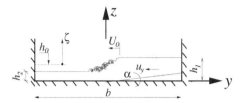

Fig. 14. Definition sketch.

capture off- and near-resonance free-surface water waves while the forcing amplitude are $a_h/b = 0.006$ or $a_h/b = 0.02$. Initially, the free-surface elevation and the velocity field are prescribed to be zero. The non-dimensional horizontal forcing parameter $\kappa_h = a_h \omega_{h_0}^2/g$ and the Froude number $Fr = U_0/\sqrt{g h_1}$ (where U_0 is the propagating speed and h_1 is the developed water depth) are used as a measure of non-linearity and bore strength, respectively.

The numerical results are compared with experimental data, as described elsewhere (Frandsen (2006a)). We also compare the LB results with a low-order Riemann solution. Herein we use a first-order upwind method with simple linear interpolations, as described by Toro (1999). We realize this is a shortcoming in our comparison study as it is well known that schemes which are first-order accurate in space and time can produce artificial viscosity. Indeed, this was experienced. Recognized shock-capturing schemes typically contain high-order interpolations with second-order accuracy methods including limiters. Extensive theories of these schemes exist in terms of flux functions (e.g. Roe, Godunov, Van Leer, Steger-Warming, etc.), reconstruction methods and limiters, however, this will not be further discussed here. Instead, the reader is referred to the comprehensive descriptions of the host of Riemann solvers in the books by Toro (2001); LeVeque (2002); and Chapter 2 in this volume by LeVeque.

All results are presented in dimensionless form. The free-surface (ζ) is non-dimensionalized by the still water depth (h_0) and the non-dimensional time $T = \omega_1 t$ and the non-dimensional time step $\Delta T = \omega_1 \Delta t$ where the first initial linear natural sloshing frequency in shallow water is defined as $\omega_1 = \sqrt{g (\pi/b)^2 h_0}$. The first numerical tests carried out were designed to check the sensitivity of the LBGK scheme to the time step and the grid resolution (Frandsen, 2006a). The 1-D simulations are based on $D_1 Q_3$ lattices. The bore formation was found to occur at $Fr_1 = 0.93$, showing two distinct developed water depths $h_1/h_2 = 1.34$. All grid resolutions used yielded

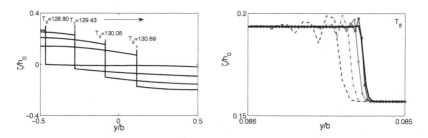

Fig. 15. LB wave profiles ($Fr_1 = 0.93$; $h_1/h_2 = 1.34$) near resonance ($\omega_h/\omega_1 = 1$; $\kappa_h = 0.003$). Grid resolution, nodes: $--$, 20,000 ($\tau=0.547$); $-\cdot-$, 30,000 ($\tau=0.578$); $-+-$, 40,000 ($\tau=0.591$); $-\circ-$, 50,000 ($\tau=0.643$); $-$, 60,000 ($\tau=0.553$).

similar wave profiles and thus represent grid converged solutions except at the bore tip, as shown in Fig. 15. Higher grid resolution is required for the near resonance cases compared to outside resonance, as expected. It was found that the LB solutions for $\kappa_h = 0.003$ required a minimum of about 10,000 nodes outside resonance and 20,000 nodes near resonance, respectively, to capture the physics experienced experimentally. However, in order to represent the bore tip adequately, this would not yield accurate solutions. The LBGK and Riemann solutions agree fairly well in terms of the free-surface elevation and velocity predictions. However, a phase lag exists which appears to be more pronounced in the wave profiles than in the time evolutions at various locations in the tank. The phase lag is caused by numerical viscosity related to the first-order upwind scheme of the Riemann solution. The low-order Riemann solution produced a slower bore propagation compared to the LBGK bores. We found that, to completely suppress the bore tip oscillations, a 60,000 node resolution for the standard LBGK model would be required (Fig. 15). The first-order Riemann solution would require about a four times smaller time step to achieve equivalent accuracy. Further, it should be noted that the reason for the requirement of the high resolution is due to the vertical jump in the surface. Lesser resolution would yield a sloped jump. Moreover, the Riemann solution with 2,000 nodes would not be able to capture a bore whereas the LB solution would. In general, the low-order Riemann solution would have a tendency to predict sloped jumps whereas the LB solution would have vertical jumps for the same grid resolution (analogous to conservative versus non-conservative methods). However, the LB bore front showed numerical oscillations at the tip (Fig. 15) which could be suppressed with an increase in resolution, or

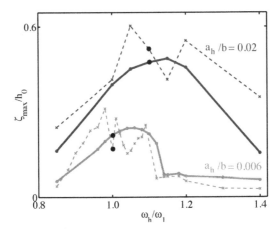

Fig. 16. Maximum free-surface elevation at the wave gauge 2 $(x, y) = (0, -b/4)$ location. —, LB solution; – –, Experimental solution. •, time evolution comparison.

perhaps better using a high-order LB finite-difference scheme. We further compare the LBGK solutions with the experimental tests series. The test series are summarized in Fig. 16 showing comparisons of the maximum free-surface elevation of the time series at forcing frequency ratio $\omega_h/\omega_1 \in [0.85, 1.2]$. The numerical near resonance test series are shown with distinct data points for the two forcing amplitudes. Figure 16 represents surface locations at the quarter locations. Fairly good agreement is achieved away from the wall whereas, at the wall itself, large discrepancies exist (not shown) due to the limitations of the numerical formulation. Furthermore, it can also be observed that the water motion exhibits a hardening spring oscillator behavior in shallow water, as the maxima occurs at $\omega_h/\omega_1 > 1$. The hysteresis effect becomes more pronounced with increasing κ, as expected. In these comparison studies, the LB solutions are based on 40,000 nodes (τ=0.0.591) and $\Delta T = 8.8 \times 10^{-5}$ whereas the Riemann solutions have 40,000 nodes and $\Delta T = 4.4 \times 10^{-5}$. Figures 17 and 18 show a comparison of the free-surface elevation for the two forcing amplitudes. The close-up look ($T \in [190, 210]$) represents steady state evolutions. The near resonance cases showed small amplitude waves in a first sloshing mode initially ($T < 30$), both experimentally and numerically. The experimental and numerical horizontal (sway) base excitation test series captured similar weak bore features. However, the forcing amplitude $a_h/b = 0.006$ required numerically to generate a weak bore was smaller than the experimental

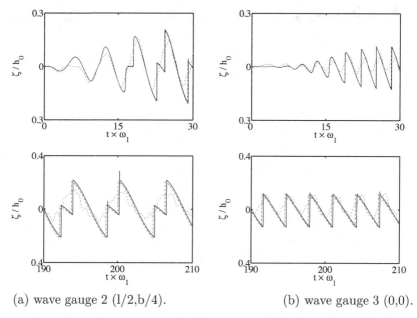

Fig. 17. Small forcing $a_h/b = 0.006$. Comparisons of free-surface evolutions at transient and steady state phases. —, LB solution; – · –, Riemann solution; – –, experimental solution.

equivalent ($a_h/b = 0.02$). It was found that the experimental tank required greater initial momentum than the numerical tank, simply occurring in the physical process of activating the movement of the tank. Therefore, it is not possible to compare test series for the exact same forcing amplitudes. Although the dominating horizontal flow maintained its overall one-dimensionality inside the tank, away from the bore tongue and walls, discrepancies were found at the bore location and at the wall due to vertical components present in the experimental data. This could also be a reason the bore is developed for smaller forcing amplitudes numerically. Further, it should be noted that the skewness of the numerical free-surface is not typical. It is caused by the bore which occurred periodically in the tank. The numerical simulations predicted a bore ($Fr_1 = 0.93$, $h_1/h_2 = 1.34$) whereas the physical experiments captured traveling waves ($Fr_0 = 1.0$, $h_1/h_2 \approx 1$). Although the numerical solutions agree overall fairly well with the experimental data, as shown in the close-up look (e.g. Fig. 18), it should be emphasized that the vertical velocity components occurring at the walls and at the location of the bore are not captured by the governing

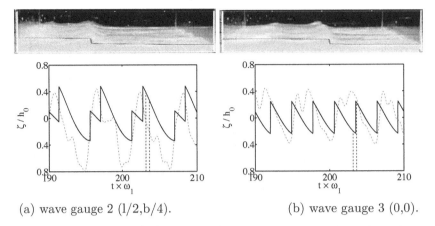

(a) wave gauge 2 (l/2,b/4). (b) wave gauge 3 (0,0).

Fig. 18. Moderate forcing $a_h/b = 0.02$. Free-surface evolutions and snapshots of wave profile comparisons (time instances labelled with vertical dashed lines). —, LB solution; — · —, Riemann solution; — —, experimental solution.

equations, and are thus a source of error inherently built into the numerical formulation. Further details and discussions on the 1-D solution can be found in Frandsen (2006a). Two-dimensional simulations with sway base excitations have also been undertaken. We found that the 1-D and 2-D weak bore solution compares fairly well.

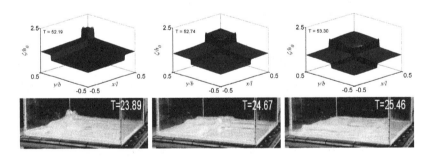

Fig. 19. Snapshots of wave profile comparisons for sway and surge base excitation.

Our third numerical test to be presented involves sway and surge base excitation for $a_{hx} = a_{hy} = 0.006$ (expanding the above pure sway case). The fluid flow is characterized by D_2Q_9 lattices. These preliminary test series include low resolution grids 500×500 ($\Delta t \times \omega_1 = 0.002$). Figure 19 shows numerically and experimentally generated time instances of the propagating bores. Two consecutive bores were observed to form, propagating diagonally at $Fr_1=1.78$ ($h_1/h_2=2.27$). The bores were also observed to form at $a_h/b=0.006$ similar to the numerical experiments. It should be noted that similar observations have been reported by Wu et al. (1998) who used a fully nonlinear 3-D potential flow solver.

The 1-D results presented herein have been computed on a single processor of 3.4 GHz. The near resonance cases were the most intensive with average LB CPU time in the order of 6 h for 60,000 node resolutions. The combined sway and surge test cases took about 8 h (low resolution grids compared to the pure sway cases).

7. Case Study: Dam-Break Simulations

The following test series represents a classical dam-break problem on a wet bed. The present test case is carried out to test the solver's further ability to handle these kind of shocks. The model is discretized with D_2Q_9 lattices (Fig. 20(b)).

The domain size is $l \times b = 200 \times 200$ m^2. The initial still water level is $h_1=10$ m when $y < b/2$ and $h_2=5$ m when $y > b/2$. A flat smooth bed is assumed and the shear stress is assumed negligible. The outer tank walls are modeled with a slip condition whereas the dam-break wall has no-slip

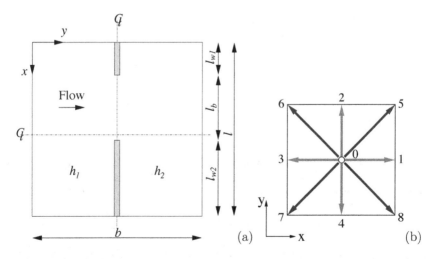

Fig. 20. (a) Definition sketch of the dam-break test case. (b) D_2Q_9 velocity set.

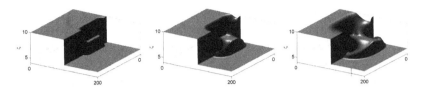

Fig. 21. Snapshots of dam-break surface flow through asymmetric wall configuration at 1.6 s, 4.9 s, 7.2 s.

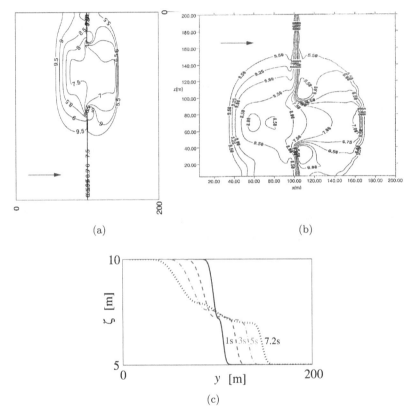

Fig. 22. Snapshots of the free-surface elevation contours of dam-break flow through the asymmetric wall configuration (in meters). (a) LBGK, (b) BGK, after Ghidaoui et al. (2001), (c) LBGK free-surface profiles at selected instances ($x = l_{w1} + l_b/2$).

conditions prescribed. The standard first-order accurate bounce-back technique is applied. The walls are located at $y = b/2$. The wall lengths l_{w1} and l_{w2} are 30 m and 95 m, respectively. The preliminary test case is set-up with a grid resolution of 400×400 and the time step is Δt=0.0025 s. The asymmetric dam-break problem has a breach $l_b = 75$ m, as shown in Fig. 20. The dam is assumed to break instantaneously. The Froude number for this study is approximately $Fr \approx 0.34$. The free-surface time evolutions of the dam-break are shown in Fig. 21. The magnitude of the elevations are shown in Fig. 22. Good agreement is found with similar studies of Fennema and Chaudhry (1990) and Ghidaoui *et al.* (2001) who used a finite discretization of the shallow water equations and a finite-volume BGK scheme, respectively. Future studies shall investigate supercritical flows in which an order of magnitude higher grid resolutions are expected if using the standard LBGK scheme. Values of h_1/h_2 larger than 2 also would require higher resolution and/or scheme improvement over the present reported test case. In summary, the LBGK appears to yield promising results. It is notable that the resolution for the dam-break test case is an order of magnitude less compared to the bore studies presented in previous test cases. We are currently undertaking further investigations.

We have done a variety of other test cases in the step-by-step implementation and validation process of the LBGK solver aiding towards modeling the tsunami-generated runup on sloped beaches. Some of the work is still

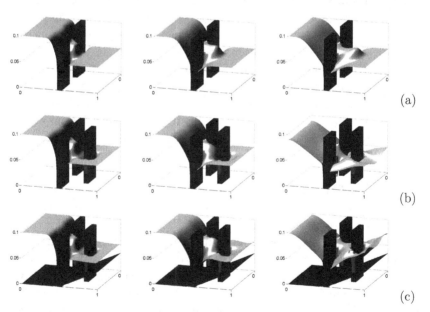

Fig. 23. (a) Dam-break problem including (b) wall movement and (c) bed slope ($\alpha = 4$); 5 s between snapshots.

in progress. However, we would like to present some preliminary test cases to the reader just to provide some insight into what would be possible with this type of solver. For example, we briefly looked at comparing the effects of adding bluff-bodies to the above dam-break problem, as shown in Fig. 23. No-slip boundaries are prescribed at the surfaces of the bluff-bodies. We also included a bed slope α ($F_x = 0; F_y \neq 0$ in Eq. (27)). The force term $F_y = -g h \frac{\partial h_b}{\partial y}$ where h_b is the height of the slope. Obviously many issues need to be verified here, especially high Fr and boundary layer physics, but overall the physics of the flow generated is promising.

8. Conclusions and Future Directions

We have investigated an LB scheme with the simplest approximation to the collision integral. We have explored the standard LBGK scheme for single phase flows with application to free-surface water waves. For incompressible Navier-Stokes equations, the scheme is second-order accurate in space and first-order accurate in time.

The 1-D and 2-D LBGK numerical solutions presented approximate the NLSW in rotational flows and can be shown to satisfy the depth-integrated Navier-Stokes equations. Fairly good agreement with Riemann solutions and model-scale experiments have been achieved for weak bore propagation in tanks. Dam-break simulations were also undertaken. The free-surface predictions compared fairly well with other investigators. The tsunami runup problem (benchmark 1) was also simulated, applying two LB schemes (1) the standard LBGK and (2) a second-order Finite Difference (FD) LB solvers. We found fairly good agreement with the NLSW solutions of Pedersen (Chapter 17 in this volume) and the semi-analytical solution of Carrier et al. (2003) when we used a shoreline algorithm of Lynett et al. (2002). The FD LB solver is, however, recommended as it most accurate and efficient for the free-surface solutions presented in the benchmark test 1 study herein. The other test cases presented in this chapter are based on the standard LBGK scheme only. In future studies,we recommend a comparison with the FD LB scheme and expect major improvements in terms of accuracy, stability and efficiency, similar to what we observed for the wave runup problem. The CPU time for all tests was competitive, especially the FD LB scheme for tsunami predictions. There are, however, many issues which can be improved. First, we recommend that a multi-time-relaxation scheme be applied instead of the single-time relaxation scheme as it would provide the stability, accuracy and improve the description of the details of the physics of the flow processes. Second, instead of the regular standard lattice scheme one could test the LB performance using adaptive irregular lattices.

Based on our current knowledge, we highly recommend that that a second-order FD LB scheme form the basis for free-surface water-wave developments. We are currently developing an LB model applicable at any water depth. This involves an entirely new LB model formulation which satisfies the Navier-Stokes equations. Our focus is on breaking wave predictions and we are investigating free-surface algorithms which will allow for steep and overturning waves. Closely related, we are also undertaking detailed bubble break-up studies based on soft-sphere collision models. An LB sediment-transport model is also part of the LB development. At some point in time, we should be able to simulate the wave transformation from deep to shallow water to runup.

On a final note, it is our hope that this book chapter represents a useful starting point for those interested in learning about and/or applying LB models. In our opinion, LB simulations are a promising route in calculation methods and may be a helpful method in discovering new physics.

Acknowledgments

I wish to thank Professor P. L.-F. Liu, Cornell University (USA), for giving me the opportunity to write about the first time experiences with the Lattice Boltzmann generated water waves. Thanks are also due to Professor G. Pedersen, University of Oslo (Norway), who provided the NLSW and Boussinesq data allowing us to assess the Lattice Boltzmann solutions in detail. I would also like to mention that I met Professor H. Yeh, Oregon State University, in Spring 2004, who actually introduced me to the wave runup problem. In many ways, this was a turning point for the Lattice Boltzmann model development and other new research directions. I am also thankful to Professor C. Dalton, University of Houston (USA), for his many valuable comments and corrections to the manuscript.

Further thanks to the many students, including under-represented minority and early career researchers, who got partly involved in this project of LB model learning. Especially, I would like to thank visiting scholar, Dr. J. Zhang, University of Hawai'i Manoā (UHM), who assisted producing the results for the benchmark test case 1 at the final phase of the chapter development. Finally, many thanks to the organizers of the faculty ambassadors program of UHM, for giving me the opportunity to share with middle and high school student the importance of practical experiences, research and education in ocean and coastal engineering including the present tsunami application and studies.

The first thoughts of exploring an LB approach to ocean waves were compiled in proposals to the US National Science Foundation (NSF) in Summer 2003, and sponsored Fall 2004 by NSF, the Division of Civil, Mechanical and Manufacturing Innovation (grant no. CMMI-CAREER-0625423) and the Division of Chemical, Bioengineering, Environmental, and Transport Systems (grant no. CBET-0622388).

Finally, but not least, the opportunity to participate in the Catalina Island NSF workshop (2004) was in so many ways extremely valuable to me. It was a highly efficient, educational and well planned meeting, thanks to the organizers, Professors P. L.-F. Liu, H. Yeh and C. Synolakis.

I am very grateful that all these happenings came my way.

References

Aidun, C. K., Lu, Y. and Ding, E.-J. (1998) Direct analysis of particulate suspensions with inertia using discrete Boltzmann equation, *Journal of Fluid Mechanics* **373**, pp. 287–311.

Benzi, R., Succi, S. and Vergassola, M. (1992) The Lattice Boltzmann equation: theory and applications, *Physics Reports (review section of Physics Letters)* **222**(3), pp. 145–197.

Bhatnagar, P., Gross, E. and Krook, M. A. (1954) A model for collision processes in gases i: small amplitude process in charged and neutral one component systems, *Physics Review A* **94**, pp. 511–526.

Boltzmann, L. (1964) *Lectures on Gas Theory* (Dover Publications).

Buick, J. M. and Greated, C. A. (2003) Lattice Boltzmann modelling of interfacial gravity waves, *Physics of Fluids* **10**(6), pp. 1490–1511.

Burgers, J. M. (1969) *Flow Equations for Composite Gases* (Academic Press).

Cao, N., Chen, S., Jin, S. and Martinez, D. (1997) Physical symmetry and lattice symmetry in the Lattice Boltzmann method, *Physical Review E* **55**(1), pp. R21–R24.

Carrier, G. F., Wu, T. T. and Yeh, H. (2003) Tsunami run-up and draw-down on a plane beach, *Journal of Fluid Mechanics* **475**, pp. 79–99.

Cercignani, C. (1988) *The Boltzmann Equation and its Applications* (Springer-Verlag).

Chapman, S. (1916) On the law of distribution of molecular velocities, and on the theory of viscosity and thermal conduction, in a non-uniform simple monatomic gas, *Philosophical Transactions Royal Society A* **216**, pp. 279–348.

Chapman, S. (1918) On the kinetic theory of gas. part ii: a composite monatomic gas: diffusion, viscosity, and thermal conduction, *Philosophical Transactions Royal Society A* **217**, pp. 115–197.

Chapman, S. and Cowling, T. G. (1958) *The Mathematical Theory of Non-Uniform Gases* (Cambridge University Press).

Chen, H., Chen, S. and Matthaeus, W. H. (1992) Recovery of the Navier-Stokes equations using a lattice-gas Boltzmann method, *Physical Review A* **45**(8), pp. R5339–R5342.

Chen, H., I. Saroselsky, I., Orszag, S. and Succi, S. (2004) Expanded analogy between boltzmann kinetic theory of fluids and turbulence, *Journal of Fluid Mechanics* **519**, pp. 301–314.

Chen, S. and Doolen, G. D. (1998) Lattice Boltzmann method for fluid flows, *Annual Review of Fluid Mechanics* **30**(4), pp. 329–364.

Crouse, B., Rank, E., Krafczyk, M. and Tölke, J. (2003) A LB-based approach for adaptive flow simulations, *International Journal of Modern Physics B* **17**(1 & 2), pp. 109–112.

Crouse, M., B. Krafczyk, Kühner, S., Rank, E. and Van Treeck, C. (2002) Indoor air flow analysis based on Lattice Boltzmann methods, *Energy and Buildings* **34**, pp. 941–949.

Dalrymple, R. A. and Rogers, B. (2006) A note on wave celerities on compressible fluid, in *Proceedings, 30th International Conference on Coastal Engineering* (San Diego, USA).

Dellar, P. J. (2001) Bulk and shear viscosities in Lattice Boltzmann equations, *Physics Review E* **64**, 031203.

Dellar, P. J. (2002) Non-hydrodynamic modes and *a priori* construction of shallow water lattice boltzmann equations, *Physics Review E* **65**, 036309.

Deng, J. Q., Ghidaoui, M. S., Gray, W. G. and Xu, K. (2001) A Boltzmann-based mesoscopic model for contaminant transport in flow systems, *Advances in Water Resources* **24**, pp. 531–550.

Denniston, C., Marenduzzo, D., Orlandini, E. and Yeomans, J. M. (2004) Lattice-Boltzmann algorithm for three-dimensional liquid crystal hydrodynamics, *Philosophical Transactions, Royal Society of London Series A Mathematical Physical and Engineering Sciences* **362**(1821), pp. 1745–1754.

Derksen, J. J. (2005) Simulations of confined turbulent vortex flow, *Computers and Fluids* **34**(3), pp. 301–318.

D'Humières, D., Ginzburg, I., Krafczyk, M., Lallemand, P. and Luo, L.-S. (2002) Multiple-relaxation-time for the Lattice Boltzmann models in three dimensions, *Philosophical Transactions, Royal Society London A.* **360**(1), pp. 437–451.

Ding, E.-J. and Aidun, C. K. (2003) Extension of the Lattice Boltzmann method for direct simulation of suspended particles near contact, *Journal of Statistical Physics* **112**(314), pp. 685–708.

Ding, E.-J. and Aidun, C. K. (2006) Cluster size distribution and scaling for spherical particles and red blood cells in pressure-driven flows at small reynolds number, *Physical Review Letters* **96**, 204502.

Enskog, D. (1921) The numerical calculation of phenomena in fairly dense gases, *Arkiv Mat. Astr. Fys.* **16**(1).

Fennema, R. J. and Chaudhry, M. H. (1990) Explicit methods for 2-D transient free-surface flows, *Journal of Hydraulic Engineering* **116**(8), pp. 1013–1034.

Filippova, O. and Hänel, D. (1997) Lattice Bolzmann simulation of gas-particle flow in filters, *Computers and Fluids* **26**(7), pp. 697–712.

Frandsen, J. B. (2006a) Free surface water wave 1-D LBGK predictions, *International Journal of Computational Fluid Dynamics* **20**(6), pp. 427–437.

Frandsen, J. B. (2006b) A Lattice Boltzmann bluff body model for VIV suppression, in *Proceedings, 25th International Offshore Mechanics & Arctic Engrg., Paper no. 92271* (Hamburg, Germany).

Frandsen, J. B. (2008) "A 1-D Lattice Boltzmann model applied to tsunami runup on a plane beach," in *Advanced Numerical Models for Simulating Tsunami Waves and Runup*, Advances in Coastal and Ocean Engineering, Vol. 10, eds. Liu, P. L.-F., Yeh, H. and Synolakis, C. (World Scientific Publishing Co.).

Geller, S., Krafczyk, M., Tölke, J., Turek, S. and Hron, J. (2006) Benchmark computations based on Lattice-Boltzmann, finite element and finite volume methods for laminar flows, *Computers and Fluids* **35**(8), pp. 888–897.

Ghidaoui, M. S., Deng, J. Q., Gray, W. G. and Xu, K. (2001) A Boltzmann based model for open channel flows, *International Journal for Numerical Methods in Fluids* **35**, pp. 449–494.

Ghidaoui, M. S., Kolyshkin, A. A., Liang, J. H., , Chan, F., Li, Q. and Xu, K. (2006) Linear and nonlinear analysis of shallow wakes, *Journal of Fluid Mechanics* **548**, pp. 309–340.

Ginzburg, I. and D'Humières, D. (2003) Multireflection boundary conditions for the lattice Bolzmann models, *Physical Review E 066614* **68**(1), pp. 1–30.

Ginzburg, I. and Steiner, K. (2003) Lattice Boltzmann model for free-surface flow and its application to filling process in casting, *Journal of Computational Physics* **185**, pp. 61–99.

Gombosi, T. I. (1994) *Gaskinetic Theory* (Cambridge University Press).

Gou, Z., Liu, H., Luo, L.-S. and Xu, K. (in press) A comparative study of the LBE and GKS methods for 2D near incompressible flows, *Journal of Computational Physics*.

Grad, H. (1949) On kinetic theory of rarefied gases, *Communications on Pure and Applied Mathematics* **2**, pp. 331–407.

Harris, S. (2004) *An Introduction to the Theory of the Boltzmann Equation* (Dover Publications).

He, X., Chen, S. and Zhang, R. (1999) A Lattice Boltzmann scheme for incompressible multiphase flow and its application in simulation of Rayleigh-Taylor instability, *Journal of Computational Physics* **152**, pp. 642–663.

He, X. and Doolen, G. (1997) Lattice Boltzmann method on curvilinear coordinates system, *Journal of Computational Physics* **134**, pp. 306–315.

He, X., Luo, L.-S. and Dembo, M. (1996) Some progress in Lattice Boltzmann method. Part I. nonuniform mesh grids, *Journal of Computational Physics* **129**, pp. 357–363.

Hibberd, S. and Peregrine, D. H. (1979) Surf and run-up on a beach: a uniform bore, *Journal of Fluid Mechanics* **95**, pp. 323–345.

Hou, S., Sterling, J., Chen, S. and Doolen, G. D. (1996) A Lattice Boltzmann subgrid model for high Reynolds number flows, *Fields Institute Communications* **6**, pp. 151–165.

Inamuro, T., Ogata, S., Tajima, S. and Konishi, N. (2004) A Lattice Boltzmann method for incomressible two-phase with large density differences, *Journal of Computational Physics* **198**, pp. 628–644.

Junk, M. (2001) A finite difference interpretation of the Lattice Boltzmann method, *Numerical Methods Partial Differential Equations* **17**(4), pp. 383–402.

Junk, M., Klar, A. and Luo, L.-S. (2005) Asymptotic analysis of the Lattice Boltzmann equation, *Journal of Computational Physics* **210**, pp. 676–704.

Kalikmanov, V. I. (2001) *Statistical Physics of Fluids. Basic Concepts and Applications* (Springer).

Kandhai, D., Soll, W., Chen, S., Hoekstra, A. and Sloot, P. (2000) Finite-difference Lattice-BGK methods on nested grids, *Computer Physics Communications* **129**, pp. 100–109.

Karniadakis, G. E. and Beskok, A. (2002) *Micro Flows. Fundamentals and Simulations* (Springer-Verlag).

Keller, H. B., Levine, D. A. and Whitham, G. B. (1960) Motion of a bore over a sloping beach, *Journal of Fluid Mechanics* **7**, pp. 302–316.

Koelman, J. M. V. A. (1991) A simple Lattice Boltzmann scheme for Navier-Stokes fluid flow, *Europhysics Letters* **15**(6), pp. 603–607.

Körner, C., Pohl, T., Rüde, U., Thürey, N. and Zeiser, T. (2006). *Parallel Lattice Boltzmann Methods for CFD Applications* (Numerical Solution of Partial Differential Equations on Parallel Computers. Eds. A. M. Bruaset, A. Tveito. Lecture Notes in Computational Science and Engineering, Vol. 51, Springer).

Körner, C., Thies, H., M., T. Thürey, N. and Rüde, U. (2005) Lattice Boltzmann model for free surface flow for modeling foaming, *Journal of Statistical Physics* **121**(1-2), pp. 179–196.

Krafczyk, M., Tölke, J. and Luo, L.-S. (2003) Large-eddy simulations with a multiple relaxation time LBE model, *International Journal of Modern Physics B* **17**(1), pp. 33–39.

Krafczyk, M. and Tölke, J. (2004) Lattice-Boltzmann methods — basics and recent progress, in *NAFEMS CFD Workshop on Simulation of Complex Flows* (Germany).

Lallemand, P. and Luo, L.-S. (2000) Theory of the Lattice Boltzmann method: dispersion, dissipation, isotropy, galilean invariance, and stability, *Physical Review E* **61**, pp. 6546–6562.

Laney, C. B. (1998) *Computational Gasdynamics* (Cambridge University Press).

Lee, K., Yu, D. and Girimaji, S. S. (2006) Lattice Boltzmann DNS of decaying compressible isotropic turbulence with temperature fluctuations, *International Journal of Computational Fluid Dynamics* **20**, pp. 401–413, doi: doi:10.1080/10618560601001122, URL http://www.ingentaconnect.com/content/tandf/gcfd/2006/0000002%0/00000006/art00006.

Lee, T. and Lin, C.-L. (2005) A stable discretization of the Lattice Boltzmann equation for simulation of incomressible two-phase flows at high density ratio, *Journal of Computational Physics* **206**, pp. 16–47.

Lesieur, M., Métais, O. and Comte, P. (2005) *Large-eddy Simulations of Turbulence* (Cambridge University Press).

LeVeque, R. J. (2002) *Finite Volume Methods for Hyperbolic Problems* (Wiley).

LeVeque, R. J. and George, D. L. (2008) "High-resolution finite volume methods for the shallow water equations with bathymetry and dry states", in *Advanced Numerical Models for Simulating Tsunami Waves and Runup*, Advances in Coastal and Ocean Engineering, Vol. 10, eds. Liu, P. L.-F., Yeh, H. and Synolakis, C. (World Scientific Publishing Co.).

Li, Y., Shock, R., Zhang, R. and Chen, H. (2004) Numerical study of flow past an impulsively started cylinder by the Lattice-Boltzmann method, *Journal of Fluid Mechanics* **519**, pp. 273–300.

Liang, J. H., Ghidaoui, J. Q., M. S. Deng and Gray, W. G. (2007) A Boltzmann-based finite volume algorithm for surface water flows on cells of arbitrary shapes, *Journal of Hydraulic Research* **45**(2).

Lockard, D. P., Luo, L.-S., Milder, S. D. and Singer, B. A. (2002) Evaluation of PowerFLOW for aerodynamic applications, *Journal of Statistical Physics* **107**(1/2), pp. 423–478.

Luo, L.-S. (2000) The lattice-gas and Lattice Boltzmann methods: Past, present, and future, in *Proceedings of the International Conference on Applied Computational Fluid Dynamics* (Beijing, China), pp. 52–83.

Luo, L.-S., Qi, D. and Wang, L. P. (2002) *Applications of the Lattice Boltzmann method to complex and turbulent flows* (Eds. M. Breuer, F. Durst, and C. Zenger. Lecture Notes in Computational Science and Engineering, Vol. 21, pp. 123-130.).

Lynett, P. J., Wu, T.-R. and Liu, P. L.-F. (2002) Modeling wave runup with depth-integrated equations, *Coastal Engineering* **46**, pp. 89–107.

McCracken, M. E. (2004) *Development and Evaluation of the Lattice Boltzmann Models for Investigations of Liquid Break-up*, PhD Thesis, Purdue University, USA.

McNamara, G. R. and Zanetti, G. (1988) Use of the Boltzmann equation to simulate lattice gas automata, *Physics Review Letter* **61**(20), pp. 2332–2335.

Mei, C. C. (1992) *The Applied Dynamics of Ocean Surface Waves* (World Scientific Publishing Co.).

Mei, R., Luo, L.-S. and Shyy, W. (1999) An accurate curved boundary treatment in the Lattice Bolzmann method, *Journal of Computational Physics* **155**, pp. 307–330.

Mei, R., Shyy, W., Yu, D. and Luo, L.-S. (2000) Lattice Bolzmann method for 3-D flows with curved boundaries, *Journal of Computational Physics* **161**, pp. 680–699.

Noble, R., Chen, S., Georgiadis, J. and Buckius, R. (1995) A consistent hydrodynamic boundary condition for the Lattice Boltzmann method, *Physics of Fluids* **7**(1), pp. 203–209.

Nourgaliev, R. R., Dinh, D. N., Theofanous, T. G. and Joseph, D. (2003) The Lattice Boltzmann equation method: theoretical interpretation, numerics and implications, *International Journal of Multiphase Flows* **29**, pp. 117–169.

Pan, C., Luo, L.-S. and Miller, C. (2006) An evaluation of Lattice Boltzmann schemes for porous medium flow simulation, *Computer and Fluids* **35**(8/9), pp. 898–909.

Pedersen, G. (2008) "A Lagrangian model applied to runup problems", in *Advanced Numerical Models for Simulating Tsunami Waves and Runup*, Advances in Coastal and Ocean Engineering, Vol. 10, eds. Liu, P. L.-F., Yeh, H. and Synolakis, C. (World Scientific Publishing Co.).

Peng, G., Xi, H. and Duncan, C. (1999) Lattice Boltzmann scheme method on unstructured grids: further developments, *Physical Review E* **59**(4), pp. 4675–4682.

Penrose, O. (1970) *Foundations of Statistical Mechanics. A Deductive Treatment* (Dover Publications).

Poularikas, A. D. (1996) *The Transforms and Applications Handbook* (CRC Press).

Premnath, K. N. (2004) *Lattice Boltzmann Models for Simulations of Drop-Drop Collisions*, PhD Thesis, Purdue University, USA.

Premnath, K. N. and Abraham, J. (2005) Simulations of binary drop collisions with a multi-relaxation-time Lattice Boltzmann model, *Physics of Fluids* **17**, p. 122105.

Premnath, K. N. and Abraham, J. (2007) Three-dimensional multi-relaxation-time (mrt) Lattice Boltzmann models for multiphase flow, *Journal of Computational Physics* **224**, pp. 539–559.

Qian, Y. H., D'Humières, D. and Lallemand, P. (1992) Lattice BGK models for Navier-Stokes equation, *Europhysics letters* **17**(6), pp. 479–484.

Que, Y.-T. and Xu, K. (2006) The numerical study of roll-waves in inclined open channels and solitary wave run-up, *International Journal for numerical methods in Fluids* **50**, pp. 1003–1027.

Riabouchinsky, D. (1932) *Sur l'analogie hydraulique des movements d'un fluide compressible.* (Institute de France, Academie des Sciences, Comptes Rendus, **195**(998–999)).

Rivet, J.-P. and Boon, J. P. (2001) *Lattice Gas Hydrodynamics* (Cambridge University Press).

Rothman, D. H. and Zaleski, S. (1997) *Lattice-Gas Cellular Automata. Simple Models of Complex Hydrodynamics* (Cambridge University Press).

Salmon, R. (1999a) The Lattice Boltzmann method as a basis for ocean circulation modeling, *Journal of Marine Research* **57**, pp. 503–535.

Salmon, R. (1999b) Lattice Boltzmann solutions of the three-dimensional planetary geostrophic equations, *Journal of Marine Research* **57**, pp. 847–884.

Seta, T. and Takahashi, R. (2002) Numerical stability analysis of FDLBM, *Journal of Statistical Physics* **107**(1/2), pp. 557–572.

Shen, M. C. and Meyer, R. E. (1963) Climb of a bore on a beach. part 3: Run-up, *Journal of Fluid Mechanics* **16**, pp. 113–125.

Sofonea, V. and Sekerka, R. F. (2003) Viscosity of finite difference lattice Boltzmann models, *Journal of Computational Physics* **184**, pp. 422–434.
Sterling, J. D. and Chen, S. (1996) Stability analysis of Lattice Boltzmann methods, *Journal of Computational Fluid Dynamics* **123**, pp. 196–206.
Stoker, J. J. (1957) *Water Waves. The Mathematical Theory with Applications* (John Wiley & Sons).
Struchtrup, H. (2005) *Macroscopic Transport Equations for Rarefied Gas Flows. Approximation Methods in Kinetic Theory* (Springer-Verlag).
Succi, S. (2001) *The Lattice Boltzmann Equation for Fluid Dynamics and Beyond* (Oxford University Press).
Sukop, M. C. and Thorne, D. T. (2006) *Lattice Boltzmann Modeling* (Springer).
Synolakis, C. E. (1987) The run-up of solitary waves, *Journal of Fluid Mechanics* **185**, pp. 523–545.
Thömmes, G., Seaïd, M. and Banda, M. K. (2007) Lattice Boltzmann methods for shallow water flow applications, *International Journal for Numerical Methods in Fluids* **55**(7), pp. 673–692.
Thürey, N. (2006) *Physically Based Animation of Free Surface Flows with the Lattice Boltzmann Method*, PhD Thesis, Der Technischen Fakultät der Universität Erlangen-Nürnberg, Germany.
Thürey, N. and Rüde, U. (2004) Free surface lattice-boltzmann fluid simulations with and without level sets, in *Proceedings, Vision Modeling and Visualization* (Stanford, USA).
Thürey, N., Rüde, U., Öchsner, M. and Körner, C. (2006a) Optimization and stabilization of LBM free surface flow simulations using adaptive parameterization, *Computers and Fluids* **35**, pp. 934–939.
Thürey, N., Rüde, U. and Stamminger, M. (2006b) Animation of open water phenomena with coupled shallow water and free surface simulations, in *Proceedings, Eurographics/ACM SIGGRAPH Symposium on Computer Animation*.
Tölke, J., Freudiger, S. and Krafczyk, M. (2006) An adaptive scheme using hierarchical grids for Lattice Boltzmann multi-phase flow simulations, *Computers and Fluids* **35**(8), pp. 820–830.
Tölke, J., Krafczyk, M., Schulz, M. and Rank, E. (1998) Implicit discretization and nonuniform mesh refinement approaches for FD discretizations of LBGK models, *International Journal of Modern Physics C* **9**(8), pp. 1143–1157.
Toro, E. F. (1999) *Riemann Solvers and Numerical Methods for Fluid Dynamics. A Practical Introduction* (Springer, p.168–171).
Toro, E. F. (2001) *Shock-Capturing Methods for Free-surface Shallow Flows* (Wiley).
Ubertini, S., Bella, G. and Succi, S. (2003) Lattice Boltzmann scheme method on unstructured grids: Further developments, *Physical Review E* **68**, 016701, pp. 1–10.
Van der Sman, R. G. M. (2006) Finite Boltzmann schemes, *Computer and Fluids* **35**, pp. 849–854.

Van der Sman, R. G. M. and Ernst, M. H. (2000) Convection-diffusion Lattice Boltzmann scheme for irregular lattices, *Journal of Computational Physics* **160**, pp. 766–782.

Vincenti, W. G. and Kruger, Jr., C. H. (1965) *Introduction to Physical Gas Dynamics* (Krieger Publishing Company).

Wu, G. X., Eatock Taylor, R. and Greaves, D. M. (2001) The effect of viscosity on the transient free-surface waves in a two-dimensional tank, *Journal of Engineering Mathematics* **40**, pp. 77–90.

Wu, G. X., Ma, Q. A. and Eatock Taylor, R. (1998) Numerical simulation of sloshing waves in a 3D tank based on a finite element method, *Applied Ocean Research* **20**, pp. 337–355.

Xu, K. (2002) A well-balanced gas-kinetic scheme for the shallow-water equations with source terms, *Journal of Computational Physics* **178**, pp. 533–562.

Yeh, H., Ghazali, A. and Marton, I. (1989) Experimental study of bore run-up, *Journal of Fluid Mechanics* **206**, pp. 563–578.

Yu, H. and Girimaji, S. S. (2006) Multi-block Lattice Boltzmann method: Extension to 3D and validation in turbulence, *Physica A* **362**, pp. 118–124.

Yu, H., Girimaji, S. S. and Luo, L.-S. (2005) DNS and LES of decaying isotropic turbulence with and without frame rotation using the Lattice Boltzmann method, *Journal of Computational Dynamics* **209**, pp. 599–616.

Yu, H., Luo, L.-S. and Girimaji, S. S. (2006) LES of turbulent square jet flow using and MRT Lattice Boltzmann model, *Computers and Fluids* **35**, pp. 957–965.

Zhong, L., Feng, S. and Gao, S. (2005) Wind-driven ocean circulation in shallow water Lattice Boltzmann model, *Advances in Atmospheric Sciences* **22**(3), pp. 349–358.

Zhou, J. G. (2002) A Lattice Boltzmann model for the shallow water equations with turbulence modeling, *International Journal of Modern Physics C* **13**(8), pp. 1135–1150.

Zhou, J. G. (2004) *Lattice Boltzmann Methods for Shallow Water Flows* (Springer).

Part 2
Extended Abstracts

CHAPTER 6

BENCHMARK PROBLEMS

Philip L.-F. Liu[1], Harry Yeh[2], and Costas E. Synolakis[3]

[1]*School of Civil and Environmental Engineering, Cornell University*
Ithaca, NY 14853 USA
E-mail: pll3@cornell.edu
[2]*Department of Civil Engineering, Oregon State University*
[3]*Department of Civil Engineering, University of Southern California*

Four benchmark problems were selected before the workshop so that numerical models can be compared, evaluated and discussed among the participants during the workshop. All of the benchmark-problem descriptions and necessary data were provided to the participants six months prior to the workshop. Based on the benchmark problems, each participant was asked to submit his/her written discussion prior to the workshop. The actual laboratory or physical measurements were only presented during the workshop in the same format, allowing the comparisons of predictions with measurements. All of the benchmark data are archived for download in the NEES Central Data Repository at https://central.nees.org/?projid=123.

1. Tsunami Runup onto a Plane Beach

This is a simple setup for tsunami runup modeling exercise: a uniformly sloping beach and no variation in the lateral direction, viz. a 2-D problem in the vertical plane. The initial-value-problem (IVP) technique introduced by Carrier, Wu and Yeh (*J. Fluid Mech.*, **475**, 79–99, 2003) is used to produce the benchmark data. For the benchmark problem #1, the beach slope is fixed at 1/10 and the initial free surface elevation is given (see below). The workshop participants are asked to compute and present the snapshots of the free surface and velocity profiles at $t = 160$ s, 175 s, and 220 s and describe the algorithm used to calculate the motion of the shoreline (the air-water-beach interface). Moreover, workshop participants are also required to present the temporal variations of the shoreline location and shoreline velocity from t = 100 s to 280 s

The problem statement and input data file can be found at http://www.cee.cornell.edu/longwave/data/benchmark_1.txt.

2. Tsunami Runup onto a Complex Three-Dimensional Beach

The 1993 Okushiri tsunami caused many unexpected phenomena. One of them was the extreme runup height of 32 m that was measured near the village of Monai in Okushiri Island. This tsunami runup mark was discovered at the tip of a very narrow gulley within a small cove. This benchmark problem is an 1/400 scale laboratory experiment of the Monai runup, using a large-scale tank (205 m long, 6 m deep, 3.4 m wide) at Central Research Institute for Electric Power Industry (CRIEPI) in Abiko, Japan. The figures shown below are the bathymetry (Fig. 1) and coastal topography (Fig. 2) used in the laboratory experiment. The incident wave from offshore, at the water depth d = 13.5 cm will be given. Note that there are reflective vertical sidewalls at y = 0, and 3.5 m as indicated in bold lines in Fig. 2(a). The primary theme of this benchmark problem is the temporal and spatial variations of the shoreline location, as well as the temporal variations of the water-surface variations at one or two specified nearshore locations. Although it was not required, animated presentation for the numerical model was encouraged.

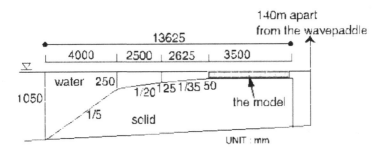

Fig. 1. Offshore profile.

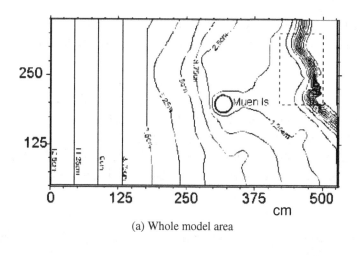

(a) Whole model area

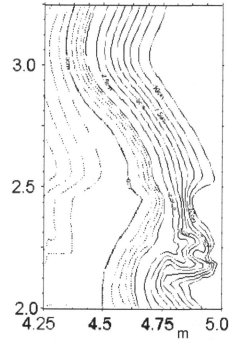

(b) Detailed topography near the maximum runup

Fig. 2. Bathymetry and topography reproduced in the tank.

The problem statement, data file, input file, bathymetry file and the experimental data file can be found in the following websites, respectively:

http://www.cee.cornell.edu/longwave/data/problem02.doc
http://www.cee.cornell.edu/longwave/data/Benchmark_2.txt
http://www.cee.cornell.edu/longwave/data/Benchmark_2_input.txt
http://www.cee.cornell.edu/longwave/data/Benchmark_2_Bathymetry.txt
http:// www.cee.cornell.edu/longwave/data/benchmark2/output_ch5-7 9.xls

3. Tsunami Generation and Runup Due to a Two-Dimensional Landslide

For the second benchmark problem, the workshop participants are asked to predict the free surface elevation and runup associated with a translating Gaussian shaped mass, initially at the shoreline. In dimensional form, the seafloor can be described by:

$$h(x,t) = H(x) + h_o(x,t),$$

where

$$H(x) = x\tan\beta,$$

$$h_o = \delta\exp\left[-\left(2\sqrt{\frac{x\mu^2}{\delta\tan\beta}} - \sqrt{\frac{g}{\delta}}\mu t\right)^2\right],$$

δ = maximum vertical slide thickness, μ = thickness/slide length, and β is the beach slope. Once in motion, the mass moves at constant acceleration. The initial position of the block is shown in the figure below which shows the start of the motion.

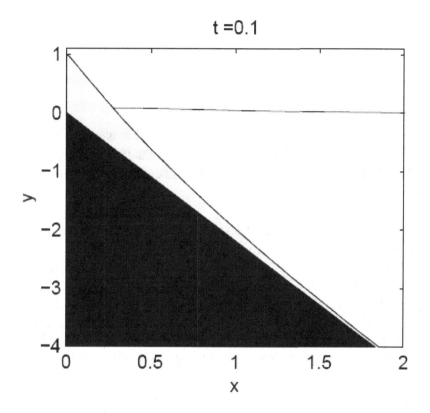

The workshop participants are asked to provide snapshots of the free surface at selected times. The following two setups will be benchmarked:

CASE A: $\tan \beta/\mu = 10 \rightarrow \beta = 5.7°, \delta = 1$ m, $\mu = 0.01$
CASE B: $\tan \beta/\mu = 1 \rightarrow \beta = 5.7°, \delta = 1$ m, $\mu = 0.1$

Four comparisons should be made for each setup. Spatial snapshots of the free surface are given at four different non-dimensional times:

CASE A: $\sqrt{\dfrac{g}{\delta}}\mu t = 0.1, 0.5, 1.0, 1.5$

CASE B: $\sqrt{\dfrac{g}{\delta}}\mu t = 0.5, 1.0, 2.5, 4.5$

The data files for this benchmark problem are named as readme1.txt, bench3A_t05.txt, bench3B_t10.txt, bench3A_t15.txt, bench3B_t05.txt, bench3B_t10.txt, bench3B_t25.txt, and bench3B_t45.txt; they can be found in the following websites:

http://www.cee.cornell.edu/longwave/data/benchmark3/readme1.txt
http://www.cee.cornell.edu/longwave/data/benchmark3/bench3A_t05.txt
http://www.cee.cornell.edu/longwave/data/benchmark3/bench3A_t10.txt
http://www.cee.cornell.edu/longwave/data/benchmark3/bench3A_t15.txt
http://www.cee.cornell.edu/longwave/data/benchmark3/bench3B_t05.txt
http://www.cee.cornell.edu/longwave/data/benchmark3/bench3B_t10.txt
http://www.cee.cornell.edu/longwave/data/benchmark3/bench3B_t25.txt
http://www.cee.cornell.edu/longwave/data/benchmark3/bench3B_t45.txt

Included in each data file are four columns of data: x (m); H (m); h_o (m); ζ (m). The free surface elevations, ζ, are those predicted by an analytical solution. Agreement with CASE B using any model other than the linear shallow water wave equations may be poor. For more information, please refer to Liu, Lynett and Synolakis (*J. Fluid Mech.*, **478**, 101–109, 2003).

4. Tsunami Generation and Runup Due to a Three-Dimensional Landslide

This benchmark problem requires the modeling of a sliding mass down a 1:2 plane beach slope and compares the predictions with laboratory data. Large-scale experiments have been conducted in a wave tank with a length 104 m, width 3.7 m, depth 4.6 m and with a plane slope (1:2) located at one end of the tank. A solid wedge was used to model the landslide. The triangular face has the following dimensions: a horizontal length of $b = 91$ cm, a vertical face $a = 46$ cm high and a width of $w = 61$ cm. The wedge was instrumented with an accelerometer to accurately define the acceleration-time history and a position indicator to independently determine the velocity- and position-time histories. The wedge traveled down the slope by gravity rolling on specially designed wheels (with low friction bearings) riding on aluminum strips with shallow grooves inset into the slope. A snapshot of the wedge motion is shown in the figure below.

A sufficient number of wave gages were used to determine the seaward propagating waves, the waves propagating to either side of the wedge, and for the submerged case, the water surface-time history over the wedge. In addition, the time history of the runup on the slope was accurately measured. The workshop participants are asked to model the flow with the wedge starting from two different initial elevations: one submerged the other subaerial. The recorded block motions are provided. The workshop participants are asked to produce time histories of surface elevation at selected locations in the channel and the shoreline motion. An animation of the results is encouraged.

The data files for this benchmark problem are named as problem statement, Landslide setup, time history (run30), time history (run32); they can be found at following websites:

http://www.cee.cornell.edu/longwave/data/Benchmark_4_readme.txt
http://www.cee.cornell.edu/longwave/data/Benchmark_4_landslide_setup.pdf
http://www.cee.cornell.edu/longwave/data/Benchmark_4_run30.txt
http://www.cee.cornell.edu/longwave/data/Benchmark_4_run32.txt

The wave gage data and runup data can be located in the following websites:

http://www.cee.cornell.edu/longwave/data/benchmark4/gage_data_readme.txt
http://www.cee.cornell.edu/longwave/data/benchmark4/run30_wave_gage_1.txt
http://www.cee.cornell.edu/longwave/data/benchmark4/run30_wave_gage_2.txt
http://www.cee.cornell.edu/longwave/data/benchmark4/run30_runup_gage_2.txt
http://www.cee.cornell.edu/longwave/data/benchmark4/run30_runup_gage_3.txt
http://www.cee.cornell.edu/longwave/data/benchmark4/run32_wave_gage_1.txt
http://www.cee.cornell.edu/longwave/data/benchmark4/run32_wave_gage_2.txt
http://www.cee.cornell.edu/longwave/data/benchmark4/run32_runup_gage_2.txt
http://www.cee.cornell.edu/longwave/data/benchmark4/run32_runup_gage_3.txt

For more information, please refer to Liu, Wu, Raichlen, Synolakis, and Borrero (*J. Fluid Mech.*, **536**, 107–144, 2005).

CHAPTER 7

TSUNAMI RUNUP ONTO A PLANE BEACH

Zygmunt Kowalik, Juan Horrillo and Edward Kornkven

Institute of Marine Science, University of Alaska Fairbanks
Fairbanks, AK 99775, USA
E-mail: ffzk@ims.uaf.edu

The problem description and setup of Benchmark Problem 1 (BM1) can be found in Chapter 6. A detailed analytical solution of the problem is described in Carrier *et al.*[1]

To solve BM1 three approaches have been carried out:

(1) First-order approximation in time
(2) Second-order approximation in time (leap-frog)
(3) Full Navier-Stokes (FNS) approximation aided by the Volume of Fluid (VOF) method to track the free surface.

Approaches (1) and (2) use one-dimensional nonlinear shallow water (NLSW) wave theory. The finite difference solution of equation of motion and the continuity are solved on a staggered grid. Both methods have second-order approximation in space. The FNS-VOF approach has been used to visualize differences against the NLSW approaches and analytical solution. The FNS equation includes the vertical component of velocity/acceleration and some differences are expected. This method solves two-dimensional transient incompressible fluid flow with free surface. The finite difference solution of the incompressible FNS equations are obtained on a rectilinear mesh.

1. Brief Description of the Methods and their Numerical Schemes

1.1. *First-Order Method*

Equations of motion and continuity read

$$\frac{\partial u}{\partial t} + u\frac{\partial u}{\partial x} + g\frac{\partial \zeta}{\partial x} + \frac{1}{\rho D}ru|u| = 0 \quad \text{and} \quad \frac{\partial \zeta}{\partial t} + \frac{\partial uD}{\partial x} = 0, \qquad (1)$$

where ρ is the water density, u is the vertically averaged particle velocity, ζ is the sea level, $D = (\zeta + H)$ is the total depth, H is the mean water depth, r is the friction coefficient and g is the gravity acceleration. The numerical solution is usually searched by using the one-time-level numerical scheme, Kowalik and Murty.[7] The numerical scheme is constructed as follows (Kowalik and Murty[8]):

$$u_j^{m+1} = u_j^m - \frac{gT}{h}(\zeta_j^m - \zeta_{j-1}^m) - \frac{u_p^m T}{h}(u_j^m - u_{j-1}^m)$$

$$- \frac{u_n^m T}{h}(u_{j+1}^m - u_j^m) \quad \frac{2T}{\rho(D_{j-1}^m + D_j^m)} r u^m |u^m|, \qquad (2)$$

$$\zeta_j^{m+1} = \zeta_j^m - \frac{T}{h}(flux_{j+1}^{m+1} - flux_j^{m+1}),$$

where $flux_j = u_p^{m+1}\zeta_{j-1}^m + u_n^{m+1}\zeta_j^m + u_j^{m+1}\frac{(H_j + H_{j-1})}{2}$, $u_p = 0.5(u_j + |u_j|)$ and $u_n = 0.5(u_j - |u_j|)$. T is the time step, h is the space step. Indices m and $j = 1, 2, 3, ...n-1$ stand for the time stepping and horizontal coordinate points, respectively. For the runup condition, the following steps are taken when the dry point $(j_{wet} - 1)$ is located to the left of the wet point j_{wet}, thus: if $(\zeta^m(j_{wet}) > -H(j_{wet} - 1))$, then $u_{j_{wet}}^m = u_{j_{wet}+1}^m$, see Kowalik and Murty.[8]

1.2. Leap Frog Method

Equation of motion and continuity are expressed in flux form as

$$\frac{\partial M}{\partial t} + \frac{\partial M^2}{\partial x} + gD\frac{\partial \zeta}{\partial x} + \frac{gn^2}{D^{7/3}}M|M| = 0 \quad \text{and} \quad \frac{\partial \zeta}{\partial t} + \frac{\partial M}{\partial x} = 0, \qquad (3)$$

where $M = uD$ is the water transport and n is the Manning's roughness coefficient. The numerical scheme is constructed on a space-time staggered grid having second-order of approximation in space and time, see Imamura et al.[4] The two-time-level numerical scheme reads

$$\zeta_j^m = \zeta_j^{m-1} - \frac{T}{h}(M_j^{m-1/2} - M_{j-1}^{m-1/2}),$$

$$M_j^{m+1/2} = \frac{1}{(1+\mu_x)}\bigg[(1-\mu_x)M_j^{m-1/2} - \frac{gD_rT}{h}(\zeta_{j+1}^m - \zeta_j^m)$$

$$- \frac{T}{h}\bigg(\lambda_1 \frac{(M_{j+1}^{m-1/2})^2}{DM_{j+1}^{m-1/2}} + \lambda_2 \frac{(M_j^{m-1/2})^2}{DM_j^{m-1/2}} + \lambda_3 \frac{(M_{j-1}^{m-1/2})^2}{DM_{j-1}^{m-1/2}}\bigg)\bigg], \qquad (4)$$

where μ_x is a friction term factor, D_M is the total depth at M points, and D_r is the total depth which depends of the sea level and depth of the neighboring cells. Parameters λ_1, λ_2 and λ_3 are the up-down wind's switches used in the nonlinear term. μ_x and D_M are defined as $\mu_x = \frac{gn^2 T}{2D_r^{7/3}}|M_j^{m-1/2}|$ and $D_{Mj}^{m-1/2} = \frac{1}{4}(D_j^{m-1} + D_{j+1}^{m-1} + D_j^m + D_{j+1}^m)$, respectively.

1.3. Full Navier-Stokes Approximation and VOF Method

Equation of continuity for incompressible fluid and the momentum equation

$$\nabla \cdot \vec{u} = 0 \quad \text{and} \quad \frac{\partial \vec{u}}{\partial t} + (\vec{u} \cdot \nabla)\vec{u} = -\frac{1}{\rho}\nabla p + \nu \nabla^2 \vec{u} + \vec{g} \qquad (5)$$

are solved in the rectangular system of coordinates, where $\vec{u}(x,y,t)$ is the instantaneous velocity vector, ρ is the fluid density, p is the scalar pressure, ν is the kinematic viscosity, $\vec{g}$ is the acceleration due to gravity and t is the time. Solution of the equations is searched using the two-step method (Chorin[2] and Harlow and Welch[3]). The time discretization of the momentum equation is given by

$$\frac{\vec{u}^{m+1} - \vec{u}^m}{T} = -(\vec{u} \cdot \nabla)\vec{u}^m - \frac{1}{\rho^m}\nabla p^{m+1} + \nu \nabla^2 \vec{u}^m + \vec{g},$$

and it is broken up into two steps as

$$\frac{\vec{\tilde{u}} - \vec{u}^m}{T} = -(\vec{u} \cdot \nabla)\vec{u}^m + \nu \nabla^2 \vec{u}^m + \vec{g}, \qquad (6)$$

$$\frac{\vec{u}^{m+1} - \vec{\tilde{u}}}{T} = -\frac{1}{\rho^m}\nabla p^{m+1}. \qquad (7)$$

Equation (7) and the continuity equation, $\nabla \cdot \vec{u}^{m+1} = 0$, can be combined into a single equation (Poisson equation) for the solution of the pressure as

$$\nabla \cdot \left[\frac{1}{\rho^m}\nabla p^{m+1}\right] = \frac{\nabla \cdot \vec{\tilde{u}}}{T}. \qquad (8)$$

The fluid free surface is described by the discrete VOF method, see Nichols and Hirt[5] and Nichols et al.[6]

2. Discussion and Conclusions

Figures 1, 2 and 3 summarize results of BM1. The simple velocity extrapolation used in the first-order method seems to follow the shoreline evolution as prescribed by the NLSW analytical solution. Sea level and velocity profile

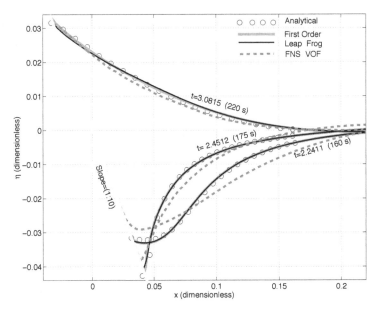

Fig. 1. Snapshots of water-surface.

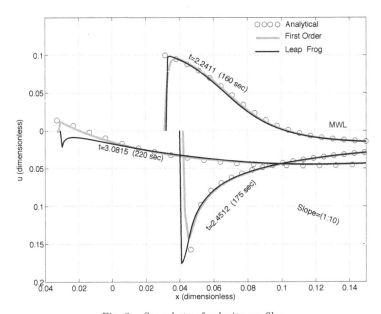

Fig. 2. Snapshots of velocity profiles.

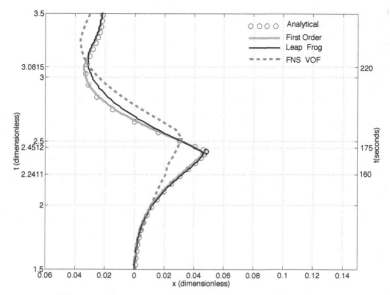

Fig. 3. Temporal and spatial variation of the shoreline.

match very well with the analytical solution. The extrapolation of the velocity from the immediate wet cell to the dry cell facilitates runup, improving the timing. The leap frog method does well in predicting the analytical solution of the shoreline evolution, sea level and velocity profile as well. However, due to the small difference in timing, some discrepancy in the velocity profile can be seen, i.e. at time $220s$. The FNS-VOF method gives a frame of reference to validate the NLSW solutions. Some differences in wave profile, shore line evolution and timing are quite plausible, since FNS approximation allows vertical fluid velocity/acceleration while the NLSW theory does not. From Figs. 1 and 3, it is clear that dispersion effects are important. NLSW and analytical solutions underestimate the runup and overestimate the rundown. Timing of maximum runup and rundown occur slightly earlier in the NLSW solutions.

References

1. G. Carrier, T. T. Wu and H. Yeh, "Tsunami runup and draw-down on a plane beach", *Journal of Fluid Mechanics*, **475**:79–99, 2003.
2. A. J. Chorin, Numerical solution of the Navier-Stokes equations, *Math. Comp.*, **22**:745–762, 1968.

3. F. H. Harlow and J. E. Welch, "Numerical calculation of time-dependent viscous incompressible flow of fluid with a free surface", *The Physics of Fluids*, **8**:2182–2189, 1965.
4. F. Imamura, C. Goto, Y. Ogawa and N. Shuto, "Numerical method of tsunami simulation with the leap-frog scheme", *IUGG/IOC Time Project Manuals*, 1995.
5. B. D. Nichols and C. W. Hirt, "Method for calculating multi-dimensional, transient free surface flow past bodies", *Proc. of the 1st Int. Conf. Num. Ship Hydrodynamics*, Gaithersburg, Maryland, 1975.
6. B. D. Nichols, C. W. Hirt and R. S. Hotchkiss, "SOLA-VOF: A solution algorithm for transient fluid flow with multiple free Boundaries", *LA-8355, Los Alamos National Laboratory*, 1980.
7. Z. Kowalik, and T. S. Murty, *Numerical Modeling of Ocean Dynamics*, World Scientific Publishing Co., 1993.
8. Z. Kowalik and T. S. Murty, "Numerical simulation of two-dimensional tsunami runup", *Marine Geodesy*, **16**:87–100, 1993.

CHAPTER 8

NONLINEAR EVOLUTION OF LONG WAVES OVER A SLOPING BEACH

Utku Kânoğlu
Department of Engineering Sciences, Middle East Technical University
Ankara, 06531, Turkey
kanoglu@metu.edu.tr

The initial value problem solution of the nonlinear shallow water-wave equations developed by Kânoğlu (2004) is applied to the benchmark problem — tsunami runup onto a plane beach — and results are compared. Comparisons with the benchmark solution produce identical results.

The solution method presented by Kânoğlu (2004) for the initial value problem solution of the nonlinear shallow water-wave equations over a sloping beach is applied to the benchmark problem. Here only the description of the method and how it is applied to the benchmark problem will be explained. Details of the method summarized here can be found in Kânoğlu (2004).

The dimensionless nonlinear shallow water-wave equations that describe a propagation problem over the undisturbed water of variable depth $h(x) = x$ are

$$[u(\eta+h)]_x + \eta_t = 0, \quad u_t + u u_x + \eta_x = 0.$$

Here $u(x, t)$, $\eta(x, t)$, and β are the horizontal depth-averaged velocity, the free-surface elevation, and the beach angle from the horizontal respectively. The initial shoreline is chosen at $x = 0$ with x increasing in the seaward-direction. The dimensionless variables are defined using $\tilde{l}$, $(\tilde{l} \tan\beta)$, and $\sqrt{\tilde{l}/(\tilde{g} \tan\beta)}$ as the characteristic length, height, and time scales; $\tilde{g}$ being the gravitational acceleration. The major analytical advance for the solution of the nonlinear shallow water-wave equations on a uniformly sloping beach was presented by Carrier and Greenspan (1958). They outlined a hodograph transformation introducing a potential $\varphi(\sigma, \lambda)$:

$$u = \frac{\varphi_\sigma}{\sigma}, \quad \eta = \frac{\varphi_\lambda}{4} - \frac{u^2}{2}, \quad x = \frac{u^2}{2} - \frac{\varphi_\lambda}{4} + \frac{\sigma^2}{16}, \quad t = u - \frac{\lambda}{2},$$

and reduced the nonlinear shallow water-wave equations into the second-order linear equation: $\sigma \varphi_{\lambda\lambda} - (\sigma \varphi_\sigma)_\sigma = 0$. The instantaneous shoreline is defined at $\sigma = 0$ in the transform space with this transformation. Given initial condition in terms of an initial wave profile with zero initial velocity, Carrier and Greenspan (1958) provided the following solution:

$$\varphi(\sigma, \lambda) = -\int_0^\infty \int_0^\infty (1/\omega)\, \zeta^2\, J_0(\omega\sigma) \sin(\omega\lambda)\, J_1(\omega\zeta)\, \phi(\zeta)\, d\omega\, d\zeta,$$

where the initial condition implies that $\phi(\sigma) = u_\lambda(\sigma, 0) = 4\eta_\sigma(\sigma, 0)/\sigma$. The major difficulty with this transformation is to drive an equivalent initial condition over the transform (σ, λ)-space for a given initial wave profile in the physical (x, t)-space. Therefore, Carrier and Greenspan (1958) presented solutions for two very specific initial profiles.

Unresolved difficulty to drive an equivalent initial condition over (σ, λ)-space for a given initial wave profile in (x, t)-space is overcome by simply with the linearization of the transformation for the spatial variable as described by Kânoğlu (2004), i.e. $x \approx \sigma^2/16$. This linearized transformation can be used to represent any initial waveform over the transform space. Then the nonlinear evolution of the wave, shoreline velocity, and shoreline runup–rundown motion can be evaluated using numerical integration.

Carrier et al. (2003) developed the Green function representation of the solution of the nonlinear shallow water-wave equations using a slightly different transformation than Carrier and Greenspan (1958). Carrier et al. (2003) evaluated Green function explicitly and obtained the highly singular complete elliptic integral of the first kind. Then, they solved the nonlinear propagation problem for arbitrary initial waveform employing numerical integration.

The initial waveform used as the benchmark problem is given by Carrier *et al.* (2003) as

$$\eta(x,t=0) = 0.006\, e^{\left[-0.4444(x-4.1209)^2\right]} - 0.018\, e^{\left[-4.0(x-1.6384)^2\right]},$$

in dimensionless form. This initial profile is used to obtain the solution for benchmark problem through Kânoğlu (2004). However, the results are converted to the dimensional quantities using a reference length $\tilde{l} = 5000$ m and a beach slope $\tan\beta = 1/10$ to compare with the benchmark solution. Figure 1 compares the initial wave profile resulted from the present nonlinear solution with the initial profile given in the benchmark problem. The comparison shows that the linearization of the spatial variable in the definition of initial condition does produce identical result. The spatial variation of the water surface elevations and velocities are compared with the benchmark solution at three different times in Fig. 2 and comparisons are excellent. In Fig. 3, comparisons are presented for the shoreline position and velocity showing identical results.

A new initial value problem solution to the nonlinear shallow water-wave equations is developed by Kânoğlu (2004) using the direct integration — without resorting to singular elliptic integrals — and is applied to the benchmark problem. Given an initial condition in the

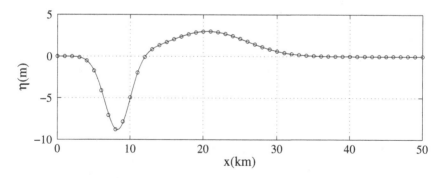

Fig. 1. Comparison of the initial wave profiles resulted from the present nonlinear solution and given in the benchmark problem. The circles represent the initial wave profile given in the benchmark problem.

physical space, the derivation of the equivalent initial condition in the transform space is possible by using linearized form of the hodograph transformation for the spatial variable at initial time. Comparisons show that direct integration yields identical results with the benchmark solution and the proposed analysis appears simpler than Carrier *et al.* (2003).

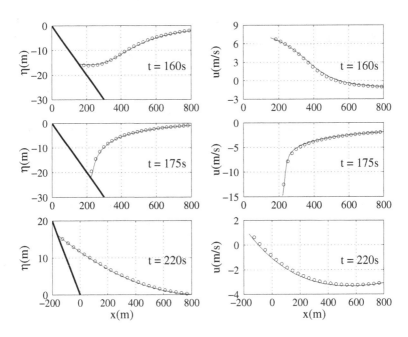

Fig. 2. Comparison of the present nonlinear solution and benchmark solution for the spatial variation of the surface elevations and velocities at three specific times. The circles represent the benchmark solution.

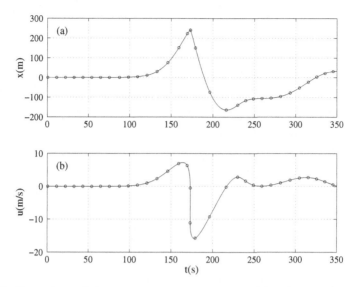

Fig. 3. Comparison of the present nonlinear solution and benchmark solution for the shoreline (a) position and (b) velocity. The circles represent the benchmark solution.

References

1. G. F. Carrier and H. P. Greenspan, *J. Fluid Mech.* **4**, 97 (1958).
2. G. F. Carrier, T. T. Wu and H. Yeh, *J. Fluid Mech.* **475**, 79 (2003).
3. U. Kânoğlu, *J. Fluid Mech.* **513**, 363 (2004).

CHAPTER 9

AMPLITUDE EVOLUTION AND RUNUP OF LONG WAVES; COMPARISON OF EXPERIMENTAL AND NUMERICAL DATA ON A 3D COMPLEX TOPOGRAPHY

Ahmet Cevdet Yalciner[1], Fumihiko Imamura[2], and Costas E. Synolakis[3]

[1]*Middle East Technical University, Civil Engineering Department, Ocean Engineering Research Center, 06531, Ankara, Turkey*
yalciner@metu.edu.tr
[2]*Tohoku University, Disaster Control Research, Aoba 980 Sendai Japan*
[3]*University of Southern California, Department of Civil and Environmental Engineering Los Angeles, 90089 LA, CA, USA*

> The runup of long waves on the sloping planes is described by the analytical solutions of the long wave equations with special initial conditions, proper approximations and boundary conditions. These studies are also compared with experimental data (Yeh *et al.* (1996); Lin *et al.* (1999); Yalciner *et al.* (2003)). Similarly the numerical methods solving governing equations with proper boundary conditions are also developed and compared with either analytical or experimental or field data for long wave propagation and runup. In this study particularly, the experimental set up is applied in the numerical model with the same wave and boundary conditions. The computed shape and amplitude evolution of the wave at the complex topography are compared with the existing experimental results. The performance of the numerical method is also discussed.

1. Introduction

Tsunami runup onto a complex 3D dimensional beach is one of the Benchmark Problems of tsunami runup. The extreme runup height of 32 m that was measured near the village of Monai in Okushiri Island by 1993 Okushiri tsunami is one of the examples of this kind of benchmark problems. There also exists experimental data in a 1/400 scale laboratory experiment of the Monai runup, using a large-scale tank (205 m long, 6 m deep, 3.4 m wide) at Central Research Institute for Electric Power Industry (CRIEPI) in Abiko, Japan. The figures showing the bathymetry, coastal topography, incident wave profile at the water depth $d = 13.5$ cm, and wave profiles at three selected locations are provided. In this study we have used the computer program TUNAMI-N2 for this problem and we compared our results of temporal and spatial variations of the shoreline location, the temporal variations of the water surface variations at three specified nearshore locations with experimental results.

2. Modeling

The numerical model, TUNAMI-N2, used for the simulation of the propagation of long waves is authored and developed by Prof. Imamura in Disaster Control Research Center in Tohoku University, Japan, through the Tsunami Inundation Modeling Exchange (TIME) program (Shuto et al. (1990)). TUNAMI-N2 is one of the key tools for the studies for propagation and coastal amplification of tsunamis in relation to different initial conditions. It solves the nonlinear form of long wave equations with bottom friction by finite difference technique, and computes water surface fluctuations and depth averaged velocities at all locations even at shallow and land regions within the limitations of grid size on the complex bathymetry and topography. Consequently the program simulates the propagation and coastal amplification of long waves. Shuto et al. (1990), Nagano et al. (1991), Imamura (1996), Goto et al. (1997), Imamura et al. (1999), Yalciner et al. (2001a, 2001b, 2002, 2003, 2004), are some of the studies used in TUNAMI-N2.

The computation domain is selected as 5.488 m by 3.402 m (393 nodes × 244 nodes with the grid size of 0.014 m). The bathymetry data is the same as the one used in laboratory experiments where the minimum depth is -0.125 m (land) and maximum depth is 0.13535 m. A typical value of 0.025 is used as Manning's coefficient (Munson et al. (1998)). The time step was chosen as 0.05 sec to satisfy the CFD condition. The propagation of the wave in the basin with respect to the inputted wave form (from laboratory data) is simulated.

3. Comparison of the Results

In order to compare the numerical and experimental results, the water surface at different times, and time histories of water surface oscillations at selected locations and the distribution of maximum positive amplitudes in the simulation duration of 22.5 sec (model time) at all grid points are computed. The computed snapshots (at $t = 12.5$ and 18.4 sec. in the model) of the water surface are shown in Fig. 1. According to simulation results, the maximum amplitude is achieved at around 18.5 sec. near the shoreline. The distribution of the computed maximum positive amplitudes in the computational domain is also shown in the same figure. According to simulation results, the maximum positive amplitude of the wave at land is found to be 0.061 m. This value is somewhat less than the maximum runup measured at the site of Monai Runup of Okushiri tsunami. One of the reasons for this difference is the local topography in the inundation zone at Monai, which could not be fully represented in the numerical model.

The measured and computed values of temporal water surface variations measured at the gauge locations (x, y) = (4.521 m, 1.196 m), (4.521 m, 1.696 m), and (4.521 m, 2.196 m) are also given in Fig. 2. The good agreement between the computed and measured water surface oscillations are obtained when the experimental and computational

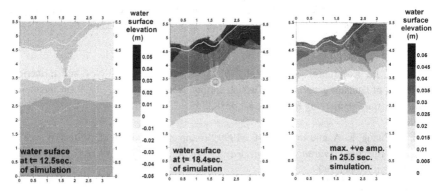

Fig. 1. The computed snapshots (at $t = 12.5$ and 18.4 sec. in the model) of the wave propagation (left two figures) and maximum positive amplitudes during 22.5 sec simulation.

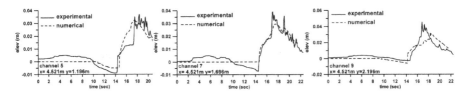

Fig. 2. The measured and computed values of temporal water surface variations measured at different gauge locations.

results for three gauge locations are compared. These results show that the simulation using TUNAMI-N2 provides fairly consistent results with experiments.

Acknowledgments

The tsunami propagation model TUNAMI-N2 is a registered copyright of Imamura, Yalciner and Synolakis.

References

Goto, C., Ogawa, Y., Shuto, N. and Imamura, F. (1997) Numerical method of tsunami simulation with the leap-frog scheme, IOC Manual 35, U.N. Educ., Sci. and Cult. Org., Paris.

Imamura, F. (1996) Review of tsunami simulation with a finite difference method, in *Long-Wave Runup Models*, eds. Yeh, H., Liu, P. L.-F. and Synolakis, C., pp. 25–42, (World Scientific Publishing Co.).

Imamura, F., Koshimura, S. and Yalciner, A. C. (1999) Field survey and numerical modeling of tsunami generated by Turkish earthquake of August 17, 1999 (in Japanese), *Proc. Coastal Engineering in Japan*, **47**, 331–333.

Lin, P., Chang, K. A. and Liu, P. L.-F. (1999) Runup and Rundown of Solitary Waves on Sloping Beaches, *Journal of Waterway, Port, Coastal, and Ocean Engineering*, September/October 1999.

Munson, B. R., Young, D. F. and Okiishi, T. H. (1998) *Fundamentals of Fluid Mechanics* (John Wiley, New York).

Nagano, O., Imamura, F. and Shuto, N. (1991) A numerical model for a far field tsunamis and its application to predict damages done to aquaculture, *Nat. Hazards*, **4**, 235–255.

Shuto, N., Goto, C. and Imamur, F. (1990) Numerical simulation as a means of warning for near field tsunamis, *Coastal Engineering in Japan*, **33**(2), 173–193.

Yalciner, A. C., Pelinovsky, E., Okal, E. and Synolakis, C. E. (Eds.) (2001a) NATO ARW, Underwater Ground Failures on Tsunami Generation, Modeling, Risk and Mitigation, North Atlantic Treaty Org., Istanbul, Turkey.

Yalciner, A. C., Synolakis, A. C., Alpar, B., Borrero, J., Altinok, Y., Imamura, F., Tinti, S., Ersoy, Ş., Kuran, U., Pamukcu, S. and Kanoglu, U. (2001b) Field Surveys and Modeling 1999 Izmit Tsunami, International Tsunami Symposium ITS 2001, Session 4, Paper 4–6, Seattle, August 7–9, 2001, pp. 557–563.

Yalciner, A. C., Alpar, B., Altinok, Y., Ozbay, I. and Imamura, F. (2002) Tsunamis in the Sea of Marmara: Historical documents for the past, models for future, *Mar. Geol.* **190**(1–2), 445–463.

Yalciner, A. C., Pelinovsky, E., Okal, E. and Synolakis, C. E. (2003) Submarine Landslides and Tsunamis, NATO Sci. Ser., Ser. IV, Vol. 21, pp. 327 (Kluwer Academic, Norwell, Mass).

Yalçiner, A. C., Pelinovsky, E., Talipova, T., Kurkin, A., Kozelkov, A. and Zaitsev A., (2004) Tsunamis in the Black Sea: Comparison of the historical, instrumental, and numerical data, *J. Geophys. Res.*, **109**, C12023, doi:10.1029/2003JC002113.

Yeh, H., Liu, P. L.-F. and Synolakis, C. (Eds.) (1996) *Long-Wave Runup Models* (World Scientific Publishing Co.).

CHAPTER 10

NUMERICAL SIMULATIONS OF TSUNAMI RUNUP ONTO A THREE-DIMENSIONAL BEACH WITH SHALLOW WATER EQUATIONS

Xiaoming Wang[*] and Phillip L.-F. Liu[†]
School of Civil and Env. Eng. Cornell University 14853, Ithaca, NY
[*]xw46@cornell.edu, [†]pll3@cornell.edu

Alejandro Orfila
IMEDEA(CSIC-UIB), Miquel Marques, 21.07190 Esporles, Spain
ao57@cornell.edu

1. Introduction

In this paper we present the numerical results for the benchmark problem #2 proposed. The model implemented is the Cornell Multigrid Coupled Tsunami Model, COMCOT (Liu et al., 1998). Although a multi-grid system, dynamically coupled up to three regions (either spherical or cartesian) can be implemented, in the present simulations only one layer has been used. The governing equations for the model are based on the shallow water wave equations. The continuity and momentum equations are expressed as,

$$\frac{\partial \eta}{\partial t} + \frac{\partial P}{\partial x} + \frac{\partial Q}{\partial y} = 0, \qquad (1)$$

$$\frac{\partial P}{\partial t} + \frac{\partial}{\partial x}\left(\frac{P^2}{H}\right) + \frac{\partial}{\partial y}\left(\frac{PQ}{H}\right) + gH\frac{\partial \eta}{\partial x} + \tau_x = 0, \qquad (2)$$

$$\frac{\partial Q}{\partial t} + \frac{\partial}{\partial x}\left(\frac{PQ}{H}\right) + \frac{\partial}{\partial y}\left(\frac{Q^2}{H}\right) - gH\frac{\partial \eta}{\partial y} + \tau_y = 0, \qquad (3)$$

where η is the free-surface displacement, h is the still water depth, $H = h+\eta$ is the total water depth and u, v are the depth averaged velocities in the

x- and y-directions. The bottom friction in equations (2)–(3) is obtained from,

$$\tau_x = \frac{gn^2}{H^{7/3}} P \left(P^2 + Q^2\right)^{1/2}, \tag{4}$$

$$\tau_y = \frac{gn^2}{H^{7/3}} Q \left(P^2 + Q^2\right)^{1/2}, \tag{5}$$

where n is the roughness coefficient. The model solves (1)–(3) with an explicit modified Leap-Frog finite difference scheme (Cho, 1995).

2. Numerical Results

The initial surface profile provide by the benchmark problem is set as the initial wave input in the left boundary and the wave propagation is simulated for 150 s. Solid reflective walls are implemented in the other three boundaries. Numerical results are compared with data provided at three gage locations A = (4.521,1.196); B = (4.521,1.696) and C = (4.521,2.196).

2.1. *Mesh Size, Bottom Fiction and Linear vs Nonlinear Model Sensitivity*

To test the sensitivity of the model to the grid size, results from two different meshes with resolutions of $\Delta x = 0.05$ m (393 × 244 grid points) and $\Delta x = 0.01435$ m (1098×681 grid points) are compared. In the coarse grid, the time increment has set as $dt = 0.001$ s so that the Courant number is 0.08 while in the finer, the time increment is $dt = 0.0002$ s being the Courant number 0.05. Figure 1(a) displays the numerical results for the fine grid (grey) and for the coarse grid (black) at the gage A. As observed, both configurations provide similar values for the arrival time of the leading wave as well as for the maximum amplitude.

The importance of the bottom friction (4)–(5) in the tsunami propagation is tested again with two simulations, one considering the bottom friction (with a fiction coefficient of n = 0.01) and the other without the friction terms. Results, displayed in Fig. 1(b), show that in the present case, the importance of the friction terms (grey line) is almost negligible providing similar results than those obtained without friction (black line). Since tsunami wavelengths are much larger than the mean water depth, the linear long wave theory should be valid providing a faster estimation than the non-linear equations. To test this hypothesis, finally the non-linear shallow

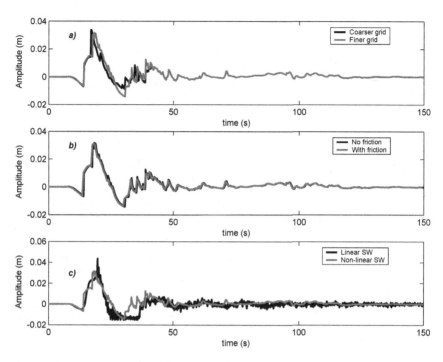

Fig. 1. Comparison between (a) grid sizes; (b) bottom friction effects and (c) linear vs. non-linear simulations.

water equations are compared with the linearized version of the equations. Figure 1(c) shows the numerical results for the linear SW equations (black) and for the nonlinear SW equations (grey) where the main differences are obtained at the small scales providing overall similar results.

3. Comparison with Gage Data

As a result from the sensitivity analysis presented in the precious section, the nonlinear model where the friction terms have been neglected is implemented and results compared with the data provided at locations A, B and C. Figure 2 displays comparison between numerical results (black) and experimental data (grey) at locations A (Fig. 2(a)), B (Fig. 2(b)) and C (Fig. 2(c)). As seen, the model provides good estimations for both the time arrival of the leading wave as well as for the amplitude of the leading and secondary waves. The nonlinear model with the coarse grid is a good

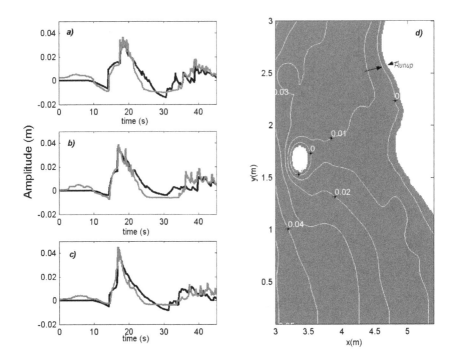

Fig. 2. (a) Free-surface elevation at gage A; (b) B and (c) C provided by the numerical model black and the experiments (grey). (d) Maximum runup superimposed to the bathymetric contours.

approximation since an agreement between grid size and time consumptions has to be considered (0.074 s per time step for the coarse grid compared to 1.193 s for the finer one). The runup, displayed in Fig. 2(d) provides a maximum value of 10.2 cm.

4. Conclusions

The numerical results show that the problem can be well simulated as a first attempt with the linear system without including bottom friction. Results indicate that the grid size reduction does not lead to a much accurate results. COMCOT results match the experimental data very good for both the arrival time and amplitude of leading waves. In the near shore region, the waves becomes very nonlinear and will break. In this zone, the present model is no longer capable to deal with it and the wave amplitude may be exaggerated.

References

1. P. L.-F. Liu, S.-B. Woo and Y. S. Cho, *Computer Program for Tsunami Propagation and Inundation*, Cornell University (1998).
2. Y.-S. Cho, *Numerical Simulations of Tsunami Propagation and Runup*, Ph.D. thesis, Cornell University (1995).

CHAPTER 11

3D NUMERICAL SIMULATION OF TSUNAMI RUNUP ONTO A COMPLEX BEACH

Taro Kakinuma

Tsunami Research Center, Port and Airport Research Institute
3-1-1 Nagase, Yokosuka, Kanagawa 239-0826, Japan
E-mail: kakinuma@pari.go.jp

This paper describes two 3D numerical models to simulate tsunami phenomena including runup onto a complex beach. The governing equations are the continuity and Reynolds equations for incompressible fluids in porous media. In the first model, water surface displacement is determined by the vertically integrated equation of continuity, while in the second model, by the 3D-VOF method. Seabed topography can be smoothly expressed with the porous model. These two models were applied to reproduce the existing hydraulic-model experiment, which treated the 1993 Hokkaido Nansei-Oki earthquake tsunami in Okushiri Island, i.e. Benchmark No. 2. The highest runup calculated by the VOF method indicates the full-scale value of about 30.6 m at the valley.

1. Introduction

From offshore to coastal zones, phenomena having various scales on space and time should be solved economically. For this purpose we have developed a hybrid model, STOC,[1] which consists of 3D models, multilevel models and connection models. As shown in Fig. 1, the multilevel models are used for wide-area calculation, where we treat hydrostatic pressure, while the connection models are applied to overlap regions for smooth connection between the multilevel models and the 3D models, where we solve pressure without assumption of hydrostatic pressure. By local application of the 3D models to narrower areas

Tab. Calculation condition.

Machine condition	
Computer	2.5GHz, desktop-type machine
Memory	930~940MB
Coding	FORTRAN 90 on LINUX

STOC-NS	
Δt	0.005~0.01s
Δx & Δy	0.014m
Δz	0.014m for $z = -0.14$~0.126m
Grid points	$393 \times 244 \times 20 = 1,917,840$
CPU time	17hrs on slip bottom for $t = 0$~25.0s, 12hrs on non-slip bottom for $t = 0$~25.0s

STOC-VF	
Δt	0.001~0.005s
Δx & Δy	0.014m
Δz	0.017~0.008m for $z = -0.15$~-0.024m, 0.008m for $z = -0.024$~0.12m
Grid points	$393 \times 244 \times 28 = 2,684,976$
CPU time	46hrs for $t = 0$~24.0s

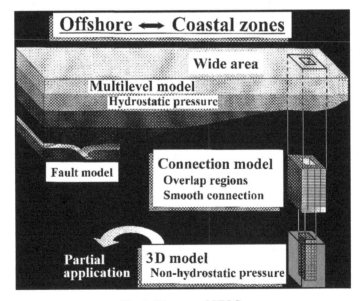

Fig. 1. Diagram of STOC.

including structures over steep topography etc., we can perform efficient and economical computation of 3D characteristics of flow, which is a blue rose for the traditional nonlinear long-wave models.

STOC includes two 3D models, i.e. STOC-NS and STOC-VF. In this paper, only these 3D models of STOC have been applied to simulate tsunami runup of Benchmark No. 2.

2. Runup Calculation Using Two 3D Numerical Simulators

The governing equations are the continuity and Reynolds equations for incompressible fluids with a porous model describing seabed topography smoothly. We solve these equations utilizing a finite difference method, which is stated in detail by Kakinuma and Tomita.[1,2]

In *STOC-NS*, we calculate water surface displacement using the vertically integrated equation of continuity. In runup regions, "Bucket-brigade method" is used, i.e. there exists water where the water depth is larger than some reference value. Unfortunately this model is not applicable to cases where the water surface elevation is described by a multi-valued function of x, because of the vertical integration.

On the other hand, *STOC-VF* is a 3D-VOF method, adopting a convection equation of the VOF function. Using this model, which can work also when the water surface elevation is described by a multi-valued function of x, we can represent breaking of water waves.

The target is to reproduce the 1/400-scale laboratory experiment[3] of the Monai runup, i.e. Benchmark No. 2. The calculation conditions are given as shown in the table. It should be noted that it is not easy to compare the CPU time for the calculation with STOC-NS and that with STOC-VF because Δt changes according to Courant number in calculation, as well as the grid-point numbers are different.

Figures 2(a)–2(c) show time variation of water level at Channels 5, 7 and 9 in Benchmark No. 2, respectively. Although the eddy viscosity is assumed to be equal to zero in STOC-VF, the results by STOC-VF are similar to those by STOC-NS, where an LES model is installed, except at Channel 9 while $t = 15 - 16.5$ (s). The reason of this difference is that STOC-NS has no wave-breaking model.

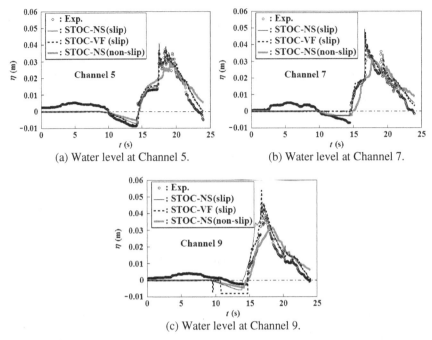

Fig. 2. Time variation of water level calculated by STOC-NS and STOC-VF in comparison with the experimental data.

As shown in Fig. 2, the calculation results of water level under the slip-bed condition show higher first-peak values than the experimental data, while those under the non-slip-bed condition show much lower water level, which suggests that some bottom-friction model is required.

After draw-down, the tsunami crest travels toward the shore, and then the maximum runup appears at Point A in Fig. 3 when $t = 16.5$ s. This runup height, calculated by STOC-VF under the slip-bed condition, is about 7.65 cm in the model scale, corresponding to 30.6 m in the prototype scale. Figure 4 shows velocity vectors and isobaric curves on the x–z plane including Point A at this time. Not only is wave height important, both fluid velocity and force are important elements in the consideration of countermeasures against tsunami disasters.

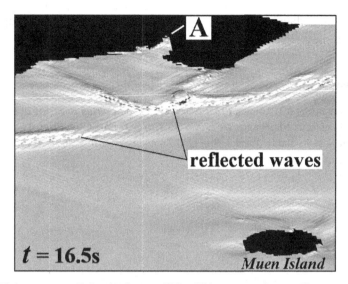

Fig. 3. Highest runup at Point A when $t = 16.5$ s. This water surface profile was calculated by STOC-VF under the slip condition.

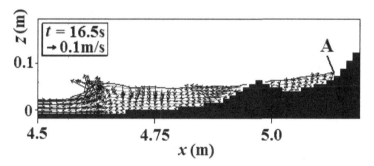

Fig. 4. Velocity vectors and constant-pressure curves on the x–z plane including Point A in Fig. 3 when $t = 16.5$ s. These were calculated by STOC-VF under the slip condition.

3. Conclusions

Two numerical, hydrodynamic models, STOC-NS and STOC-VF, have been applied to Benchmark No. 2, where water surface displacement is determined by different ways. The calculation results by both models under the slip-bed condition generally show correspondence with the experimental data, resulting in indication of about 30.6 m runup at the Monai Valley.

References

1. T. Kakinuma and T. Tomita, *Proc. 29th Int. Conf. on Coastal Eng.*, 1552 (2004).
2. T. Kakinuma and T. Tomita, *OCEANS/TECHNO-OCEAN Conf. Proc.*, 146 (2004).
3. M. Matsuyama and H. Tanaka, *Proc. Int. Tsunami Symposium 2001*, 879 (2001).

CHAPTER 12

EVALUATING WAVE PROPAGATION AND INUNDATION CHARACTERISTICS OF THE MOST TSUNAMI MODEL OVER A COMPLEX 3D BEACH

Arun Chawla[*], Jose Borrero[†] and Vasily Titov[*,‡]

Joint Institute for the Study of Atmospheres and Oceans
NOAA Center for Tsunami Research
[†]*University of Southern California*
[‡]*Vasily.Titov@noaa.gov*

1. Introduction

Runup measurements from the 1993 tsunami off the coast of Okushuri island in Japan showed that the highest wave was measured at the end of a narrow gully in a cove near the village of Monai. The measured runup mark at 32 m was significantly higher than other recorded runup values around the island, and was the largest recorded measurement in Japan over the last century.[2] To determine how the local bathymetric and topographic features could act as a focusing mechanism for wave energy, a laboratory scale experiment of this region was carried out at the Central Research Institute for Electric Power Industry (CRIEPI) in Akibo, Japan. This experimental data set provides a valuable benchmark for evaluating long-wave tsunami runup models and was part of a recent international workshop on long-wave runup models.

In this paper the laboratory data set is used to evaluate a 2D long-wave model (MOST), which is the primary tsunami inundation model used by the NOAA Center for Tsunami Research at the Pacific Marine and Engineering Laboratory in Seattle. It is currently being implemented as a forecast model at the two U.S. Tsunami Warning Centers.[4] The model has been developed by Titov and Synolakis[3] and has been validated with analytical solutions, the conical island experimental studies of Briggs *et al.*[1] and several historic tsunami events. The aim here is to characterize how accurately the model can simulate wave propagation and inundation processes in a more complex but still controlled environment.

2. Laboratory Experiments

Simulations of wave runup around the village of Monai in Okushuri island were carried out with a 1/400 scale laboratory experiment using a large scale tank (205 m long, 6 m deep, 3.4 m wide) at the Central Research Institute for Electric Power Industry (CRIEPI) in Akibo, Japan. This benchmark problem was part of the "3rd International Workshop on Long-Wave Runup Models" held on June 17–18, 2004, at the Wrigley Marine Science Center in Catalina Island California.

Figure 1 shows a side view of the experimental setup. Waves are shoaled from the deep water using a series of linearly sloping beaches. The area marked as "the model" in the figure is where the complex bathymetric and topographic features from the field have been reproduced and marks the limit of the numerical modeling exercise. The bathymetry for this model area is shown in Fig. 2. The wave propagates from left to right, with solid walls along the sides and the right end. The incoming wave is determined from a gage located near the offshore boundary and is given in Fig. 3.

3. Model

MOST is a 2D long-wave model that uses the method of fractional steps[5] to reduce the 2D shallow water wave equations into a series of 1D equations along each spatial dimension. Each set of 1D equations is re-written in characteristic form and solved using an explicit finite difference scheme (Method of undetermined coefficients). The scheme is second-order accurate in space and first-order accurate in time. Solving a series of 1D equations is computationally more efficient than solving the 2D equations. Detailed

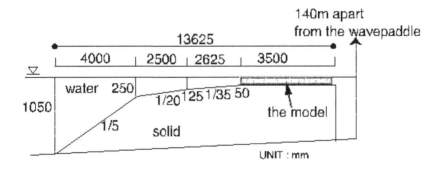

Fig. 1. Side view sketch of the experimental setup.

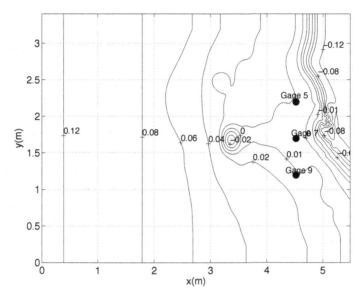

Fig. 2. Domain bathymetry in the model area. The smallest depth is -0.125 m and the deepest offshore depth is 0.135 m. Both the cove and the offshore island from the field have been reproduced in this physical model. The three gage locations used for model–data comparisons are given by the solid black circles.

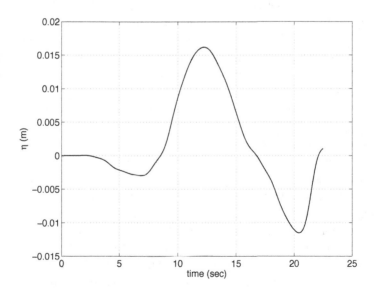

Fig. 3. Incoming wave at the boundary $x = 0$.

information on the model algorithm and numerical scheme can be obtained in Titov and Synolakis.[3]

In its current form the model has been made flexible enough to allow for multiple nested grids, and can be run in spherical or cartesian coordinate systems. There is no limit on the number of nested grids allowed and any particular grid can have a series of nested grids under it. The nested grids have the option of two-way or one-way (coarse to fine) coupling. The model also has the option of switching on wetting and drying capability in all the grids. Earlier versions of the model only allowed wetting and drying in the finest grids, thus limiting the depth to which the coarser grids could be extended in shallow water. A manning formulation is used to account for bottom friction. For these runs the manning coefficient has been set to zero (no friction). The model does not have a separate treatment for wave breaking. Breaking appears as a shock wave in the numerical solution.

4. Results

The simulations were carried out using two grids of different resolutions, with the coarser grid covering the full domain and the finer grid nested inside. The coarser grid had a resolution of $\Delta x = 0.028$ m and a time step $\Delta t = 0.02$ s. The surface time series in Fig. 3 was provided as the boundary condition along the left boundary ($x = 0$). The corresponding horizontal velocity was determined from shallow water linear wave theory. Reflecting boundary conditions were used along the remaining three boundaries. The internal grid had a resolution of $\Delta x = 0.014$ m and a time step $\Delta t = 0.01$ s. Its boundary conditions were determined from the solutions of the outer grid. Imposing boundary conditions from the coarser grid on the incoming waves and allowing the outgoing waves to radiate out are trivial with the MOST model since it solves a set of characteristic equations along each dimension. Wetting and drying was switched on for both the grids, with one-way coupling between the coarse and fine grid. The simulation was run on an a single processor of a 4 processor Intel Xeon 3.6 GHz machine running Red Hat Linux and took approximately 2.7 minutes to complete.

Time series comparison of model and data for the three wave gages in shallow water (Fig. 2) are shown in Fig. 4. The gap in the model time series at gage 9 corresponds to drying. There was a phase lag of 0.6 s between the data and the model at all the three gages. This could either be because there is an offset in the start time of the model runs as compared to the data, or that the offshore gage is not exactly at $x = 0$ as in the model. Accounting

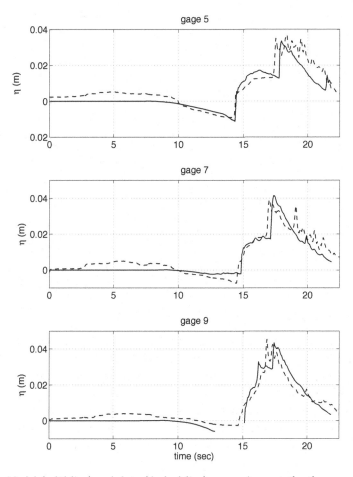

Fig. 4. Model (solid line) and data (dashed line) comparisons at the three gage locations. See Fig. 2 for gage locations.

for that phase lag, we see from the figure that the model compares very well with the data. The high frequency waves at the wave crest are not reproduced in the model simulations. Considering the shallow water depth at these gages (approximately 1 cm), the high frequency waves are probably caused by higher order details of wave breaking which are not simulated in the long wave model. Maximum alongshore (y-direction) wave runup from the simulations is shown in Fig. 5. The highest value of 10.047 cm is observed at the cove.

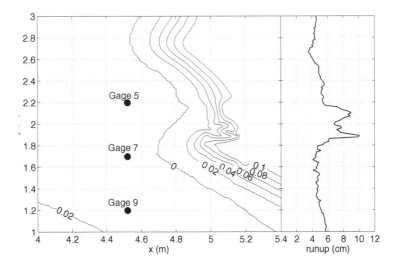

Fig. 5. Maximum wave runup along the coast.

5. Conclusions

Earlier laboratory validation studies of the MOST tsunami model[3] have shown that the model compares well in simple domains. The model is currently being used to develop a short-term tsunami forecasting capability[4] and has been considerably revised over the last year. Some of the changes include a generalized methodology for nesting grids and the capability to have wetting and drying in all the grids. These comparisons with a laboratory study in a complex domain show that the model performs reasonably well. This increases our confidence in the model.

Acknowledgments

This publication is [partially] funded by the Joint Institute for the Study of the Atmosphere and Ocean (JISAO) under NOAA Cooperative Agreement No. NA17RJ1232, Contribution number 1315.

References

1. M. J. Briggs, C. E. Synolakis, G. S. Harkins and D. R. Green, Laboratory experiments of tsunami runup on circular island, *Pure and Applied Geophys.* **144** (1995) 569–593.
2. N. Shuto and H. Matsutomi, Field survey of the 1993 Hokkaido Nansei-Oki tsunami, *Pure and Applied Geophys.* **144** (1995) 649–664.

3. V. Titov and C. Synolakis, Numerical modeling of tidal wave runup, *Waterway, Port, Coastal and Ocean Eng.* **124**(4) (1998) 157–171.
4. V. Titov, F. I. González, E. N. Bernard, M. C. Eble, H. O. Mofjeld, J. C. Newman and A. J. Venturato, Real-time tsunami forecasting: challenges and solutions, *Nat. Hazards* **35**(1), Special Issue, U.S. National Tsunami Hazard Mitigation Program (2005) 41–58.
5. N. Yanenko, *The Method of Fractional Steps*, translated by M. Holt (Springer, New York, Berlin, Heidelberg, 1971).

CHAPTER 13

TSUNAMI GENERATION AND RUNUP DUE TO A 2D LANDSLIDE

Zygmunt Kowalik, Juan Horrillo and Edward Kornkven

Institute of Marine Science, University of Alaska Fairbanks
Fairbanks, AK 99775, USA
E-mail: ffzk@ims.uaf.edu

The objective of this problem is to predict the free surface elevation and runup associated with translating a Gaussian shaped mass which is initially at the shoreline. The problem description of Benchmark Problem 3 (BM3) can be found in Chapter 6. A detailed analytical solution of the problem is described in Liu et al.[3]

Two problems are given in BM3 to be solved:

A. $\tan\beta/\mu = 10$, where: $\beta = 5.7°$, $\delta = 1$ m and $\mu = 0.01$
B. $\tan\beta/\mu = 1$, where: $\beta = 5.7°$, $\delta = 1$ m and $\mu = 0.1$,

where δ is the maximum vertical slide thickness, $\mu = \delta/L$ is the slide thickness-length ratio, and β is the slope angle.

To solve BM3 two approaches have been carried out:

(1) First-order approximation in time
(2) Full Navier-Stokes (FNS) approximation aided by the Volume of Fluid (VOF) method to track the free surface.

Approach 1 uses one-dimensional linear and nonlinear shallow water wave theories, (LSW) and (NLSW) respectively. The finite difference solution of equation of motion and the continuity equation is solved on a staggered grid, Kowalik and Murty.[4] This method has second-order approximation in space and first-order in time. More detailed information is indicated in Chapter 7.

The two-dimensional FNS-VOF solution has been included to visualize differences against the shallow water solutions. The FNS-VOF method uses the full nonlinear Navier-Stokes equation to model two-dimensional transient and incompressible fluid flow with free surfaces. The finite difference solution is obtained on a rectilinear mesh. The model has been extended to deal with rigid moving objects in the computational domain. Since FNS equation includes the vertical component of velocity/acceleration, some differences are expected if nonhydrostatic effects are strong.

1. Full Navier-Stokes Approach and VOF Method Including Moving Objects

Equation of continuity for incompressible fluid and the momentum equation,

$$\nabla \cdot \vec{u} = \frac{1}{V}\frac{dV_o}{dt} - \phi \quad \text{and} \quad \frac{\partial \vec{u}}{\partial t} + \nabla \cdot (\vec{u}\vec{u}) - \phi\vec{u} = -\frac{1}{\rho}\nabla p + \nu\nabla^2\vec{u} + \vec{g}, \quad (1)$$

are solved in the rectangular system of coordinates, where $\vec{u}(x,y,t)$ is the instantaneous velocity vector, ρ is the fluid density, p is the scalar pressure, ν is the kinematic viscosity, $\vec{g}$ is the acceleration of gravity and $\phi(\vec{x},t)$ is a moving object internal function needed to force zero divergence in the computational control volume V. The momentum and continuity equations take into account the conservation mass due to the incursion or retreat of a moving object in the domain.

Solution of the equations is searched using the two-step method (Chorin[1] and Harlow and Welch[2]). The time discretization of the equations is similar to that described in Chapter 7.

2. Discussion and Conclusions

Results obtained using first-order numerical model and LSW analytical solution are depicted in Fig. 1 for Case A. Agreements of the numerical model results (NLSW and LSW) with the analytical solution are quite good. The analytical solution does an excellent job in predicting wave runup and wave propagation for thin slide, i.e. $\tan\beta/\mu = 10$.

For thicker slide, Case B, omission of nonlinearity leads to disagreements in later stages of wave propagation, i.e. $t > 1.0$ (see Figs. 2 and 3). Note that the LSW numerical result follow very well its analytical solution.

A FNS-VOF solution is presented for case B in order to compare with the NLSW model solution. The FNS-VOF solution is ideally suited for this case, where relatively high vertical acceleration occurs. Due to dispersion effect FNS-VOF method predicts a wave a little bit more elongated, skewed but slightly less tall than the NLSW model, i.e. at $t = 2.5$ and $t = 4.5$. All numerical models and analytical solution agreed very well at earlier time (i.e. $t = 0.5$ and $t = 1$).

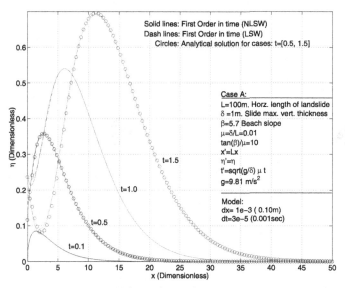

Fig. 1. CASE A: $(\tan\beta/\mu = 10)$. Free surface comparison.

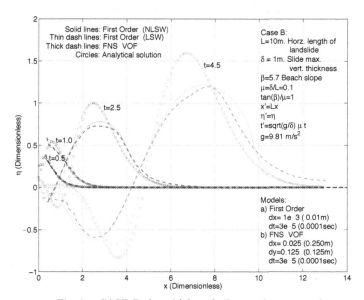

Fig. 2. CASE B: $(\tan\beta/\mu) = 1$. Free surface comparison.

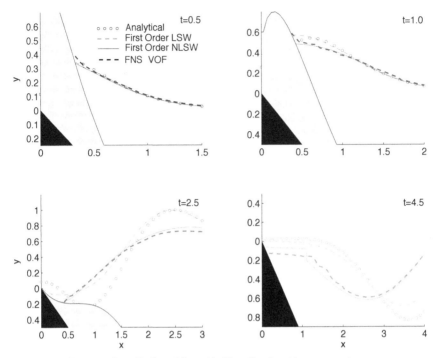

Fig. 3. Case B: $(\tan\beta/\mu = 1)$. Shoreline location comparison.

References

1. A. J. Chorin, Numerical solution of the Navier-Stokes equations, *Math. Comp.*, **22**:745–762, 1968.
2. F. H. Harlow and J. E. Welch, Numerical calculation of time-dependent viscous incompressible flow of fluid with a free surface, *The Physics of Fluids*, **8**:2182–2189, 1965.
3. P. L.-F. Liu, P. Lynett and C. E. Synolakis, Analytical solution for forced long waves on a sloping beach, *Journal of Fluid Mechanics*, **478**, 101–109, 2003.
4. Z. Kowalik and T. S. Murty, *Numerical Modeling of Ocean Dynamics*, World Scientific Publishing Co., 1993.

CHAPTER 14

BOUSSINESQ MODELING OF LANDSLIDE-GENERATED WAVES AND TSUNAMI RUNUP

Okey Nwogu

Dept. of Naval Architecture and Marine Engineering, University of Michigan
Ann Arbor, MI 48109, U.S.A.
E-mail: onwogu@umich.edu

1. Introduction

The massive destruction and loss of life associated with recent tsunamis (Indonesia, 2004; Papua New Guinea, 1998) has underscored the need to develop and implement tsunami hazard mitigation measures such as early warning systems in tsunami-prone areas. Given either the seabed deformation or ocean measurements of water surface elevation by buoys, hydrodynamic models can be used to forecast in real-time the propagation and transformation of tsunamis across the world's ocean basins and the subsequent inundation of low-lying coastal areas.

Numerical models based on the non-dispersive shallow water equations are often used to simulate tsunami propagation and runup (e.g. Titov and Synolakis [6]). The shallow water equations assume that the vertical fluid motions are much smaller than the horizontal fluid motions and the resulting fluid pressure is hydrostatic. The hydrostatic pressure assumption is reasonable for seismic tsunamis since most ocean basins have depths of O(1km) while earthquake-generated tsunamis typically have wavelengths of O(100km).

Landslide-generated tsunamis are characterized by shorter wavelengths [O(10km)], depending on the relative slide speed or depth-based Froude number. The propagation of tsunamis with water depth to wavelength ratios of $O(10^{-1})$ can be more accurately modeled by dispersive water wave evolution equations such as Boussinesq-type equations (e.g. Nwogu [3]). For dispersive waves, the horizontal velocities are no longer uniform over the depth and the pressure field is no longer hydrostatic. The velocity and pressure field is derived by asymptotically expanding the velocity potential in terms of a frequency dispersion parameter. The flow field is then substituted into the mass and momentum equations to yield Boussinesq-type equations. Boussinesq equations have also been used to simulate landslide-generated waves by modifying the boundary condition at the seabed (e.g. Lynett and Liu [2]). This paper documents the extension of a Boussinesq model to landslide tsunami problems and the validation of the model with benchmark problems as part of the Third International Workshop on Long-Wave Runup Models.

2. Description of Numerical Model

For surface gravity waves with wavelengths much longer than the characteristic water depth, the vertical profile of the velocity field can be approximated by expanding the velocity potential as a Taylor series about an arbitrary elevation z_α in the water column (e.g. Nwogu [3]). The velocity potential that satisfies the Laplace equation and moving bottom boundary condition can be written as:

$$\phi(x,y,z,t) = \phi_\alpha(x,y,t) + \mu^2(z_\alpha - z)\left[h_t + \nabla\phi_\alpha \cdot \nabla h\right]$$
$$+ \frac{1}{2}\mu^2\left[(z_\alpha + h)^2 - (z+h)^2\right]\nabla^2\phi_\alpha, \quad (1)$$

where $h(x, y, t)$ is the water depth, μ is a frequency dispersion parameter and $\phi_\alpha = \phi(x, y, z_\alpha, t)$. The above expansion can be substituted into the continuity equation and the 3-D Euler equations of motion to yield:

$$\eta_t + \nabla \cdot \left[(h+\eta)\bar{u}\right] = -h_t - \nabla \cdot \left[(h+\eta)\left\{(z_\alpha + h) - \frac{h+\eta}{2}\right\}\nabla h_t\right],$$

$$\begin{aligned}
u_{\alpha,t} &+ g\nabla \eta + (u_\eta \cdot \nabla)u_\eta + w_\eta \nabla w_\eta \\
&- \left[(u_{\alpha,t} \cdot \nabla h) + (h+\eta)\nabla \cdot u_{\alpha,t}\right]\nabla \eta \\
&+ (z_\alpha - \eta)\left[\nabla(u_{\alpha,t} \cdot \nabla h) + (\nabla \cdot u_{\alpha,t})\nabla h\right] \\
&+ \frac{1}{2}\left[(z_\alpha + h)^2 - (h+\eta)^2\right]\nabla(\nabla \cdot u_{\alpha,t}) \\
&+ \frac{1}{h+\eta}\nabla[\nu(h+\eta)\nabla \cdot u_\alpha] + \frac{1}{h+\eta}f_w u_b |u_b| = -(z_\alpha - \eta)\nabla h_{tt},
\end{aligned}$$
(2)

where $\eta(x,y,t)$ is the water surface elevation, u_α is the horizontal velocity at $z = z_\alpha$, g is the gravitational acceleration, f_w is a bottom friction coefficient, ν is a wave energy dissipation coefficient, $\bar{u}$ is the depth-averaged horizontal velocity, u_η and w_η are the horizontal and vertical velocities at the free surface ($z = \eta$), and u_b is the horizontal velocity at the seabed ($z = -h$). The different horizontal and vertical velocities are obtained from Eq. (1). The equations (2) have been numerically integrated using a finite-difference method (Nwogu and Demirbilek [4]). The moving boundary algorithm is described in [5].

3. Benchmark Problems

3.1. *Tsunami Runup onto a Complex Three-dimensional Beach*

In this benchmark problem, we attempt to reproduce the results of a physical model study of wave runup over three-dimensional topography during the 1993 Okushiri tsunami. The bathymetry is shown in Fig. 1. The complex nature of the bathymetry led to wave focusing and a maximum runup height of 32m at the tip of a cove.

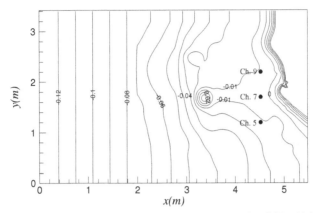

Fig. 1. 2-D map of bathymetry for laboratory model study of Okushiri tsunami.

Numerical simulations were performed with grid size $\Delta x = \Delta y = 0.04$m, and time step $\Delta t = 0.01$s. The model was initialized with the measured water surface elevation along the offshore boundary. Figure 2 shows a comparison of the measured and predicted time histories of the water surface elevation at Gauge #9. The Boussinesq model does reproduce the maximum height of the tsunami in shallow water although there are differences in the phasing of the reflected wave.

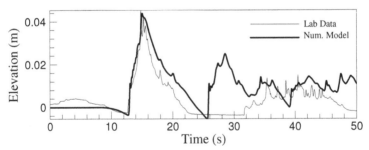

Fig. 2. Measured and predicted wave elevation time histories at Gauge #9 for Okushiri tsunami.

3.2. *Tsunami Generation and Runup Due to a Two-dimensional Landslide*

We consider the problem of a Gaussian-shaped mass sliding down a beach of slope β. The instantaneous seabed elevation is given by:

$$h(x,t) = x\tan\beta + \delta\exp\left[-\left(2\sqrt{xh_\mu^2/\delta\tan\beta} - h_\mu t\sqrt{g/\delta}\right)^2\right], \quad (3)$$

where δ is the maximum vertical slide thickness and h_μ is the thickness/slide length. Liu et al. [1] derived an analytical solution for the problem based on the linearized shallow water equations. Numerical simulations were carried out for both the nonlinear shallow water (NLSW) and Boussinesq equations for $\beta = 5.7°$, $\delta = 1$m with $h_\mu = 0.01$ (Case A) and $h_\mu = 0.1$ (Case B). For Case A, the NLSW and Boussinesq solutions were identical to the analytical solution and are not presented here. For Case B with a larger slide thickness to length ratio, differences can be seen between the three solutions as shown in Fig. 3.

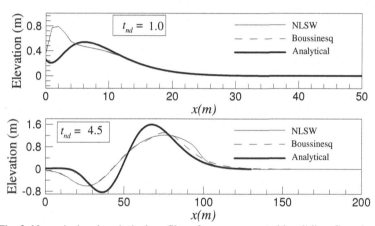

Fig. 3. Numerical and analytical profiles of waves generated by sliding Gaussian mass.

References

1. P.L.-F. Liu, P. Lynett, and C. Synolakis, *J. Fluid. Mech.*, **478**, 101 (2003).
2. P. Lynett and P.L.-F. Liu, *Proc. R. Soc. Lond.*, **458**, 2885 (2002).
3. O. Nwogu, *J. Waterw. Port Coastal Ocean Eng.*, **119**, 618 (1993).
4. O. Nwogu and Z. Demirbilek, *Tech. Rep. ERDC/CHL 01-25*, U.S. Army Engineer R & D Center, Vicksburg, MS (2001).
5. O. Nwogu and Z. Demirbilek, *Coastal Engineering*, accepted (2005).
6. V. Titov and C.E. Synolakis, *J. Waterw. Port Coastal Ocean Eng.*, **121**, 308 (1995).

CHAPTER 15

NUMERICAL SIMULATION OF TSUNAMI RUNUP ONTO A COMPLEX BEACH WITH A BOUNDARY-FITTING CELL SYSTEM

Hiroyasu Yasuda

*River Engineering Division, Civil Engineering Research Institute of Hokkaido
Hiragishi 1-3, Toyohira, Sapporo, 062-8602, Japan
E-mail: h-yasuda@ceri.go.jp*

Tsunami runup onto a complex beach was modeled and analyzed using a method that allows accurate reproduction of complex topographic features. The analysis produced accurate results and confirmed that topography greatly affects the flow regime.

1. Introduction

The tsunami caused by the 1993 Hokkaido-Nansei-Oki Earthquake off southwestern Hokkaido left traces of a great runup at Monai Beach on Okushiri Island. This runup is regarded as one of the highest of the 20th century, and its severity is attributed in part to the very complex 3-dimensional trough-shaped topography of that beach. This study reproduced the runup using 2-dimensional analysis that incorporates a "boundary-fitting cell" (BFC) system. The system allows calculations to flexibly incorporate complex topographic shapes.

2. Method of Numerical Analysis

2.1. *Dividing the Analysis Domain*

Numerical analysis was divided into two domains: the steeply undulating Monai Beach, and other locations. For discretization of topographic shapes, the former domain used BFCs to allow maximum incorporation of topographic shapes at the location, and the latter domain used a Cartesian grid.

2-dimensional analysis[1] using BFCs allows accurate and efficient incorporation of topographic shapes by being able to combine triangular and polygonal cells without restriction. An additional benefit of efficient discretization is the significantly reduced calculation time. This allows calculation that can be one hundred times faster than that using a uniformly subdivided Cartesian grid. Furthermore, it allows incorporation of linear structures, such as road or wall structures, that greatly affect the flow but that are too narrow to be incorporated as grids.

Discretization of the two steep-sloped trough and ridge topographic shapes at Monai Beach was performed by first defining the sides of the cell along the trough line and ridge line. Then the sides of the cell were defined as either parallel to or perpendicular to the elevation contour line.

2.2. Governing Equations

Two governing equations of fluid motion were applied to the Cartesian grid domain: the equations were the shallow water equation, all terms in the equation were integrated from the bottom to the water surface. The seabed frictional resistance is given by Manning's equation. One governing equation was applied in the BFC domain: It was obtained by expanding the shallow water equation in the 2-dimensional plane, ignoring the convection terms, to allow calculation using polygonal cells.

$$\frac{1}{l}\frac{\partial Q}{\partial t} + gD\frac{\partial \eta}{\partial s} = -\frac{gn^2 Q|Q|}{D^{7/3}}, \quad (1)$$

$$\frac{\partial \eta}{\partial t} - \frac{1}{A}\sum_{i=1}^{k} Q_i = 0, \quad (2)$$

where t is the time coordinate, Q is flow rate at the sides of the cell, s is the plane coordinate, η is the water level, D is the water depth, g is the gravitational acceleration, n is Manning's roughness coefficient, A is the area of the cell where the water level is to be obtained, and k is the number of sides of the cell.

The numerical scheme used in Eqs. (1) and (2) is based on the Staggered Leap-Frog scheme, which is an explicit differential scheme.

2.3. Boundary and Initial Conditions

For the Cartesian grid domain, these were the boundary conditions: the upper edge, the lower edge, and the section of the right edge with elevation of less than 0 cm were considered a fully reflective boundary. The section at the right edge with elevation of more than 0 cm was not considered a

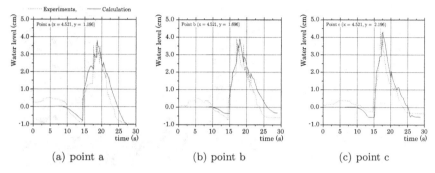

Fig. 1. Time-lapse comparison of water levels: experiments vs. numerical analysis.

reflective boundary and tsunami was able to run up onto land. The entire surface of the left edge was considered an incidence boundary and a water level was assigned. For the BFC domain, this was the boundary condition: the water levels obtained from the Cartesian grid calculation were assigned to cells at the domain boundary.

For both domains, the initial conditions were these: at 0 cm height, static water surface was assigned; and at points over 0 cm height, the water depth was set at 0 cm.

3. Analysis Results

3.1. *Cartesian Grid Domain*

The accuracy in calculating sea level variations is an indication of the validity of boundary conditions in the BFC domain. Figure 1 shows time-lapse waveforms of water levels obtained from the hydraulic experiment and numerical analysis. The three points in Fig. 1 show that the time-lapse water levels for both methods are in close agreement.

3.2. *BFC Domain*

The results of 2-dimensional tsunami runup analysis at Monai Beach using the BFCs described in §2.1 are shown in Fig. 2(b). The figure shows water depth distributions at the time of maximum runup area. The curves in this figure, from left to right, show 2.5, 5.0, and 7.5 cm elevation contours.

The BFC and Cartesian grid results generally agree in terms of flooding area and water depth distribution. There is only a slight difference of 0.2 sec in the calculated time of maximum flood area. The maximum runup

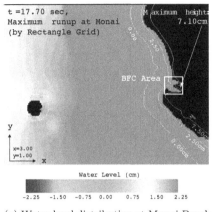 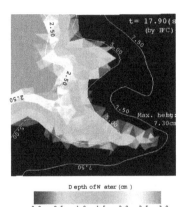

(a) Water level distribution at Monai Beach at the time of maximum flood runup

(b) Flow regime of flood from the BFC method

Fig. 2. Analysis results for the tsunami runup at Monai Beach. (a) shows the results of analysis using the Cartesian grid: the maximum runup height at Monai Beach is 7.1 cm. In BFC analysis, the water level on these results are obtained.

height calculated using BFC was expected to be higher, but it remained at 7.3 cm, just slightly higher than the 7.1 cm height calculated using the Cartesian grid. The main reason the calculated runup height did not reach the observed value, even with the BFC method, is thought to be that the convection terms is neglected in the equation of BFC. Even though inertial forces in the flow predominated, the neglect of convection terms resulted in an underestimation of runup height.

4. Conclusion

This numerical analysis calculated tsunami runup onto a complex beach, using the BFC system, which accurately reproduces complex topographic features. The calculations obtained maximum runup heights that were similar to results, without 3-dimensional calculation. This shows the importance of using discretization methods that are appropriate for the flow characteristics and topography.

References

1. H. Yasuda, M. Shirato, C. Goto and T. Yamada, *Journal of Hydraulics, Coastal and Environment Engineering* **No.740/II-64** (2003) (in Japanese).

CHAPTER 16

A 1-D LATTICE BOLTZMANN MODEL APPLIED TO TSUNAMI RUNUP ONTO A PLANE BEACH

J. B. Frandsen

*School of Civil Engineering, The University of Sydney
NSW 2006, Australia
Email: jbehrndtz@yahoo.com*

1. The Numerical Model

We consider the finite discrete-velocity model of the Boltzmann equation with a finite discrete-velocity set. The discrete Boltzmann equation with a BGK collision term, referred to as the LBGK model, as originally described by Bhatnagar *et al.* (1954), has been extended to simulate free-surface flows in shallow water.

Our model approximates the 1-D depth-averaged NonLinear Shallow Water (NLSW) equations in rotational flows,

$$\frac{\partial h}{\partial t} + \frac{\partial (h\,u)}{\partial x} = 0\,;$$
$$\frac{\partial (h\,u)}{\partial t} + \frac{\partial (h\,u^2)}{\partial x} = -g\frac{\partial}{\partial x}\left(\frac{h^2}{2}\right) + \frac{\partial}{\partial x}\left(h\,\nu\,\frac{\partial u}{\partial x}\right) + F\,,$$

(1)

where g is gravity due to acceleration, $h = h_0 + \zeta$, h_0 is the still water depth, ζ denotes the free-surface elevation measured vertically above the still water level, u is the depth-averaged horizontal velocity and ν is the kinematic viscosity. The force term, $F = -g\,h\,\frac{\partial h_b}{\partial x}$, in the present test case, accounts for the slope of the beach where h_b is the slope height, as shown in Fig. 3(a). We have not accounted for the effects of bed friction, sediment transport and wave breaking.

The numerical flow field is represented by particles which follow lattice points which stream and collide when meeting. With reference to the Boltzmann equation of classical kinetic theory, the distribution of fluid molecules

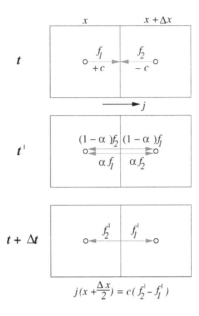

Fig. 1. D_1Q_3 collision model.

is represented by the particle distribution function $f_i(x, c_i, t)$ where i denotes propagation direction. The function defines the mass density ρ of lattice particles, equivalent to the water column h, at time t which moves with the microscopic velocity c_i. The discrete-velocity set, at a lattice node, satisfies the Maxwell distribution of equilibrium.

Herein, the numerical model is a simplification of the D_2Q_9 model approach, as shown in Chapter 5 in this volume. In the present case study, it is assumed that, at any time, the LB fluid is characterized by the populations of the three discrete microscopic velocity model, referred to as D_1Q_3; representing a 1-D model. Figure 1 shows a collision between two particles of a D_1Q_3 model. During the collision a portion α of the lattice particle collide and reverses velocity. The net flux j at the boundary between lattice cell at x and $x + \Delta x_i/2$ as

$$j(x + \Delta x_i/2) = c_i\, f'_i(x,t) + \tilde{c}_i\, \tilde{f}'_i(x + \Delta x_i, t)$$
$$= c_i\, [f_i(x + \Delta x_i, t + \Delta t) + \tilde{f}'_i(x + \Delta x_i, t + \Delta t)], \quad (2)$$

where f'_i denotes the post-collision particle distribution at time t, f_i is the pre-collision particle distribution at $t + \Delta t$, and $\tilde{c}_i$ denotes a particle propagating in the opposite direction of particles with velocity c_i ($\tilde{c}_i = -c_i$).

We introduce the single phase 1-D LBGK formulation for shallow water flows on sloped beds as

$$\frac{\partial f_i}{\partial t} + c_i \frac{\partial f_i}{\partial x} = -\frac{(f_i - f_i^{eq})}{\tau} - \frac{c_i \Delta t}{N_i c^2} g h \frac{\partial h_b}{\partial x}, \quad \text{where} \quad N_i = \frac{1}{c^2} \sum_{i=0}^{2} c_i c_i, \tag{3}$$

with N_i as a constant which depends on the lattice geometry and is defined $N_i = 2$ for the D_1Q_3 velocity set. The first term represents the effect of the local change of the fluid motion in time and the second term describes the convection. Note the convection operator of the Lattice Boltzmann Method (LBM) is linear. This property originates from kinetic theory. The first term of (3) on the right-hand-side (RHS) is the non-equilibrium distribution function which describes the effect of the collisions (Succi (2001)). The second term on the RHS represents the force term F. The time-scale parameter $\tau = 3\nu/c^2 + \Delta t/2$ where ν is the kinematic viscosity of water and $c = \Delta x/\Delta t$ (the time step is Δt and the grid spacing is Δx). It describes the collisional relaxation to the local equilibrium and $1/\tau$ is the collision frequency. In the present formulation τ is limited to a single value (BGK approximation). To ensure numerical stability $\tau > \Delta t/2$ when using the standard LBGK scheme. Further, we note that assuming a constant value of τ is considered to be a crude approximation. The equilibrium distribution functions f_i^{eq} represents the invariant function under collision (no gradients are involved) and is dependent upon the microscopic velocity c_i. It is essential that the equilibrium distribution functions satisfy the Navier-Stokes equation for any water wave model. In our case, the depth-averaged shallow-water equations with rotational flow. The f_i^{eq} functions can be derived from the Boltzmann equation using a Chapman-Enskog expansion. Details of the Chapman-Enskog method for the classical Boltzmann equation can be found in Gombosi (1994). The LBGK based f_i^{eq} expressions are velocity dependent, both at the micro and macroscale levels. A new set of equilibrium distribution functions following the 1-D flow approximation has been derived and are given as,

$$f_0^{eq} = h - \frac{g h^2}{2c^2} - \frac{h u^2}{c^2}; \quad f_{i=1,2}^{eq} = \frac{gh^2}{4c^2} + \frac{h u}{2c} c_i + \frac{h u^2}{2c^2}, \tag{4}$$

where the discrete velocity set is $c_{i=1} = 1$ and $c_{i=2} = -1$, respectively.

The hydrodynamic moments of the equilibrium distribution functions are conserved at every time step,

$$\sum_{i=0}^{2} f_i^{eq} = h \; ; \quad \sum_{i=0}^{2} c_i f_i^{eq} = hu \; ; \quad \sum_{i=0}^{2} c_i c_i f_i^{eq} = \frac{1}{2} gh^2 + hu^2 \; . \quad (5)$$

For the present application, it is noted that the pressure field exhibit near hydrostatic behavior and therefore the dynamic pressure term (hu^2) is negligible. The macroscopic variables of the free surface (ζ) and the depth averaged velocities (u) are calculated as the first and second moments of the distribution function,

$$\zeta = \sum_{i}^{2} f_i - h_0 \quad \text{and} \quad u = \frac{1}{(h_0 + \zeta)} \sum_{i}^{2} c_i f_i \; . \quad (6)$$

2. LB and the Depth-Averaged Navier-Stokes Equations

Equation (3) can be viewed as a special finite-difference discretization of the single time-relaxation approximation of the Boltzmann equation for discrete velocities. One approach to solve the discrete Boltzmann equations is to use a first-order forward Euler time-difference scheme and a first-order upwind space discretization for the convection term in a uniform lattice spacing Δx. One can then obtain following algebraic relation,

$$f_i(\vec{x} + \Delta \vec{x}, t + \Delta t) - f_i(\vec{x}, t) = -\frac{\Delta t}{\tau} \left(f_i(\vec{x}, t) - f_i^{eq}(\vec{x}, t) \right) + \frac{\Delta t}{N_i c^2} c_i F_i \; . \quad (7)$$

This equation is commonly referred to as the Lattice-BGK equation or Lattice-Boltzmann Equation (LBE). The LBE is second-order accurate in space and first-order in time (Junk et al. (2005)). See also further discussion on this matter in Chapter 5 in this volume. Further, it should be noted that using a first-order Euler scheme does not translate into the same lack of accuracy as when used in traditional CFD schemes because of the nonlinearity embedded in the collision term.

The macro dynamical behavior arising from the LBE can be found from a multiple-scale analysis such as the Chapman-Enskog technique, as described by Gombosi (1994). We shall show that for a given set of f_i (which are solutions of (3) for specific polynomials of the macroscopic quantities f_i^{eq} and moments (5)) are solutions of the depth-averaged Navier-Stokes equations. We assume small Knudsen numbers $\varepsilon = l/L$ ($\approx \Delta t$) where l is the scale in the mean free path and L is the characteristic macroscopic length. The pressure is defined as the ideal gas equation $p = c_s^2 \rho$ in flows of low Mach numbers where $c_s^2 = \partial p/\partial \rho = c^2/3$ is the speed of sound and ρ is the fluid density.

In the following we use the Chapman-Enskog technique to show that the present LB formulation recovers the NLSW equations in rotational flow. The Chapman-Enskog expansion assumes that the convection and diffusion processes operate at different time scales. The diffusion time scale being much slower than the convective counter part. First, we perform a Taylor expansion of (7) and retain the terms up to second-order,

$$\Delta t(\frac{\partial}{\partial t} + c_i \frac{\partial}{\partial x})f_i + \frac{\Delta t^2}{2}(\frac{\partial}{\partial t} + c_i \frac{\partial}{\partial x})^2 f_i = -\frac{f_i - f_i^{eq}}{\tau} + F_i \Delta t . \quad (8)$$

Then we expand the particle distribution function as

$$f_i = f_i^{eq} + \varepsilon f_i^{(1)} + \varepsilon^2 f_i^{(2)} + \cdots . \quad (9)$$

Substituting (9) into (8), the equations with the first-order and second-order of the small parameter ε are obtained as

$$\varepsilon^1 : \frac{\partial f_i^{eq}}{\partial t} + c_i \frac{\partial f_i^{eq}}{\partial x} = -\frac{f_i^{(1)}}{\tau} + F_i , \quad (10)$$

$$\varepsilon^2 : \frac{\partial f_i^{eq}}{\partial t} + c_i \frac{\partial f_i^{eq}}{\partial x} + \varepsilon\left(1 - \frac{1}{2\tau}\right) \cdot \left(\frac{\partial f_i^{(1)}}{\partial t} + c_i \frac{\partial f_i^{(1)}}{\partial x}\right)$$

$$= -\frac{f_i^{(1)} + \varepsilon f_i^{(2)}}{\tau} + F_i . \quad (11)$$

The conservation of mass and momentum requires

$$\sum_i f_i^{(n)} = 0, \sum_i f_i^{(n)} c_i = 0, n = 1, 2, \cdots . \quad (12)$$

Then summing (10) over i gives

$$\sum_i \frac{\partial f_i^{eq}}{\partial t} + \sum_i c_i \frac{\partial f_i^{eq}}{\partial x} = -\sum_i \frac{f_i^{(1)}}{\tau} + \sum_i F_i . \quad (13)$$

From (6) and (12), we obtain the macro continuity equation (1). Next, we multiply (11) by c_i and sum over i giving

$$\sum_i c_i \frac{\partial f_i^{eq}}{\partial t} + \sum_i c_i c_i \frac{\partial f_i^{eq}}{\partial x} + \varepsilon\left(1 - \frac{1}{2\tau}\right) \cdot \sum_i c_i \left(\frac{\partial f_i^{(1)}}{\partial t} + c_i \frac{\partial f_i^{(1)}}{\partial x}\right)$$

$$= -\sum_i c_i \frac{f_i^{(1)} + \varepsilon f_i^{(2)}}{\tau} + \sum_i c_i F_i . \quad (14)$$

Then we use (6), (12) and $\sum_i c_i c_i f_i^{eq} = \frac{1}{2}gh^2 + hu^2$ leading to the macro momentum equation of the NLSW equation (1),

$$\frac{\partial(hu)}{\partial t} + \frac{\partial(hu^2)}{\partial x} = -g\frac{\partial}{\partial x}(\frac{h^2}{2}) + \frac{\partial F_\nu}{\partial x} + F_i, \qquad (15)$$

where $F_\nu = \nu \cdot \left[\frac{\partial(hu)}{\partial x} + \frac{\partial(h^2/2)}{\partial t} + 3\frac{\partial(hu^2)}{\partial t}\right]$ and the kinematic viscosity is $\nu = \frac{c^2}{3}(\tau - \frac{1}{2})$.

3. Second-Order Finite Difference LB Model

We have compared the standard LB solution (7) with a second-order Finite Difference (FD) LB Model in an attempt to improve the accuracy and stability of the predictions as well as the CPU efficiency. The numerical FD scheme proposed consists of a spatial discretization of convection/collision operator, force terms and a time integration of (3). For the spatial discretization we use the second-order accurate Lax-Wendroff approximation of the convection and force terms,

$$\begin{aligned}\frac{\partial f_i(x,t)}{\partial x} &= \frac{f_i(x+\Delta x,t) - f_i(x-\Delta x,t)}{2\Delta x} \\ &\quad - \frac{\Delta t[f_i(x+\Delta x,t) - 2f_i(x,t) + f_i(x-\Delta x,t)]}{2\Delta x^2}, \\ \frac{dh_b}{dx} &= \frac{h_b(x+\Delta x,t) - h_b(x-\Delta x,t)}{2\Delta x} \\ &\quad - \frac{\Delta t[h_b(x+\Delta x,t) - 2h_b(x,t) + h_b(x-\Delta x,t)]}{2\Delta x^2},\end{aligned} \qquad (16)$$

where the force term takes the averaged value of the two values at the lattice point and its neighboring lattice point $F_i = \frac{1}{2}[F_i(x,t) + F_i(x+\Delta x, t+\Delta t)]$. The force term representing the beach slope is included in the streaming step. The discrete formulation for the convection operation is

$$C_i = c_i \cdot \nabla f_i = c_i \frac{\partial f_i}{\partial x}. \qquad (17)$$

There currently exists different suggestions to the discretization of the collision operator, e.g. Mei and Shyy (1998); Guo and Zhao (2003). For the argument of simplicity and preliminary testing, we used the suggestion as described by Mei and Shyy (1998). They proposed a simple linear extrapolation approach to calculate the f_i^{eq} at a new time level $t + \Delta t$,

$$f_i^{eq,t+\Delta t} = 2f_i^{eq,t} - f_i^{eq,t-\Delta t}. \qquad (18)$$

The collision operator $\Omega_i^{t+\Delta t}$ can be expressed as

$$\Omega_i^{t+\Delta t} = -\frac{\Delta t}{\tau}[f_i^{t+\Delta t} - (2f_i^{eq,t} - f_i^{eq,t-\Delta t})] . \quad (19)$$

The mixed numerical scheme utilizes a second-order Runge-Kutta time integrator. The time step update of the LBGK (3) reads

$$f_i^{t+\Delta t} = f_i^t + \Delta t(-C_i^{t+\Delta t/2} + \Omega_i^{t+\Delta t/2} + F_i^{t+\Delta t/2}) , \quad (20)$$

where

$$C_i^{t+\Delta t/2} = c_i \cdot \nabla f_i^{t+\Delta t/2}; \quad f_i^{t+\Delta t/2} = f_i^t + \frac{\Delta t}{2}(-C_i^t + \Omega_i^t + F_i^t)$$
$$\Omega_i^{t+\Delta t/2} = -\frac{\Delta t}{\tau}(f_i^{t+\Delta t/2} - f_i^{eq,t+\Delta t/2}) . \quad (21)$$

Substituting (17) and (19) into (20) yields the finite-difference LB equations applied herein.

4. The Free-Surface Dynamics Treatment

It is notable that the present 1-D LBGK model does not include the conventional dynamic and kinematic boundary conditions at the free-surface. Instead, the physics of the free-surface dynamics are accounted for through the non-equilibrium particle distribution function ($f^{neq} = f_i - f_i^{eq}$). So no additional surface boundary conditions are necessary to be prescribed in the present non-overturning numerical NLSW wave model. Furthermore, it is essential to satisfy the constraints of the hydrodynamics moments (5) at all times; the LB solvers equivalent means of conserving mass and momentum. We should stress that the present model can only expect to work well for a non-breaking surface in shallow water depths. Accounting for wave breaking and/or other water depths would involve a different LB formulation than the model presented herein. For example, an LB formulation in arbitrary water depth could include a free-surface treatment as known from traditional CFD solvers, e.g. a volume of fluid algorithm, a level set, etc. to account for breaking. Therefore a host of models could potentially be developed to bridge the gap between micro and macroscopic fluid flow behavior and thus potentially advance the way we currently model free-surface flows and wave runup.

5. The Wet-Dry Interface and Other Boundary Treatments

The computational domain is defined over $x \in [-500, 50500]$ m, discretized with N lattices. Beyond the beach slope, at the shore boundary end of the computational domain ($x = -500$ m), we have not prescribed any boundary condition as the water level does not reach this level in the benchmark test case 1 (see problem set-up in §7). A numerical damping zone is required to avoid non-physical reflections at the open boundary ($x = L$). In the LB method, the boundary conditions are described through distribution functions f_i. At $x = L$, we are required to calculate f_i when $i = N$. To handle reflections, we have extended the flow domain with a stretch of flat bed from $N = 50$ km to $N_d = 500$ m to mimic some form of sponge layer. We have adopted the simple extrapolation equation $f_i(N_d, t) = 2f_i(N_d - 1, t) - f_i(N_d - 2, t)$ of Yu et al. (2005) at $x = 50,500$ m. It should be noted that by imposing these constraints, the reflections are not suppressed. They are only delayed so that the present solutions in the time duration of interest are not affected. It should also be mentioned that this has not been dealt with previously, that is, in the LB free-surface framework.

Regarding the wet-dry interface treatment, the "slot method" is one technique to use in wave runup predictions, e.g. Kennedy et al. (2000). The method assumes a porous beach allowing for the water level to be below the beach level. Several other techniques were reviewed and evaluated by, e.g. Balzano (1998); Prasad and Svendsen (2003). We have tested two methods in simulating runup. The first method prescribes a thin film ($h = h_s = 10^{-5}$ m) in dry areas ($h = 0$) along the slope of the beach. The shoreline location is identified when $\zeta \to h_s$. The second method involves the use of a shoreline algorithm. We have chosen to apply the algorithm of Lynett et al. (2002). Their moving boundary technique utilizes linear extrapolation of the surface properties near the wet-dry boundary in which the shoreline is traced on a fixed grid. In the following, we shall refer to our results as predictions with or without a shoreline algorithm, respectively.

6. Solution Procedure

Initially we prescribe the waveform of Fig. 3(b) together with the non-physical values of the velocity $u = 0$ and the distribution functions $f_i = f_i^{eq}$. We solve (7) using a non-splitting operator scheme for both the standard LBGK and the FD scheme. The solution procedure is outlined in Fig. 2. The links relevant to the LB schemes herein are labeled: standard LBGK and FD LB, respectively. The hydrodynamic moments of the equilibrium distribution functions of (5) are conserved at every time step. It is assumed that the Mach number Ma $= u/c_s \ll 1$, where c_s is the speed of sound. The

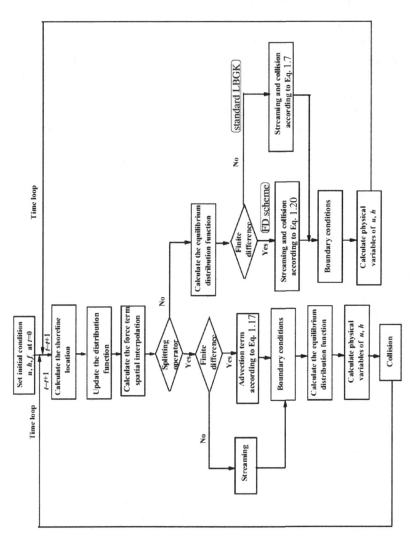

Fig. 2. Solution procedure. Non-splitting operator scheme applies to the standard LBGK and FD LB solvers.

Peclet number $Pe = u\,\Delta x/\nu < 2$ and the Courant number $Cr = u\,\Delta t/\Delta x < 1$ are also satisfied.

7. Benchmark Problem 1 Set-up

In the following test case, we are concerned with testing the solver's ability to handle wave runup on beaches. The predictions of the shoreline trajectory represent a classical bench mark test of numerical models, especially because of the challenge of accurately predicting the wave motion when the depths are vanishing into dry-states. The present test case represents a tsunami wave generated runup/run-down study on a beach. The initial shoreline location is located 50,000 m from starting point of the slope and $z_0 = 5000$ m. The slope of the beach is constant $\partial h_b/\partial x = 1/10$ and uniform and spans the whole computational domain, as shown in Fig. 3(a).

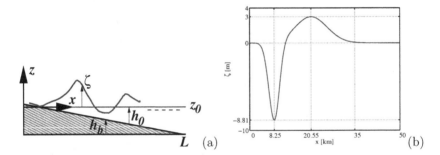

Fig. 3. (a) Definition sketch of wave runup study. (b) Initial wave profile.

Initially the velocity in the flow domain is zero and the free-surface described by the form of a leading depression N-wave shape, typically caused by an offshore submarine landslide,

$$\zeta = a_1\,e^{(-k_1(x-x_1)^2)} - a_2\,e^{(-k_2(x-x_2)^2)} \quad \text{and} \quad u = 0 \quad \text{at t=0}, \qquad (22)$$

where $a_1 = \frac{1}{3}a_2 = 0.006$, $k_1 = \frac{1}{9}k_2 = 0.4444$, $x_1 = 4.1209$ and $x_2 = 1.6384$, after Carrier et al. (2003). The initial wave profile is also shown in Fig. 3(b).

8. Results

We shall, in the following, present 1-D wave runup results based on our standard LBE and second-order accurate FD LBM solvers. First we undertook the standard LBE simulations without any shoreline algorithm. In these analysis, a thin film is prescribed to treat the wet-dry interface, as mentioned. We observed relative good comparisons regarding the inundation length (x) except for the maximum run-down location which was underestimated. The accuracy of the shoreline velocity (u_s) was poor. The solver breaks near the maximum run-down occurrence. The second-order FD LB showed similar results although the accuracy of the maximum run-down inundation length improved (and fewer nodes were required). To overcome these difficulties, other options for the wet-dry interface treatment were explored, and, as mentioned, the shoreline algorithm of Lynett et al. (2002) was implemented. The above simulations were repeated and the second-order FD LB solutions were found to yield similar results as the semi-analytical solutions of Carrier et al. (2003), as shown in Fig. 4. Using the FD solver, we need about 5,000 nodes and about 1/2 hour on a standard PC of today of 3 GHz to solve the wave runup problem of benchmark 1. This is not as fast as a standard NLSW solver but still acceptable when it comes to tsunami predictions. To support these findings, we display the behavior of the macroscopic variables in Fig. 8. First, we show the grid sensitivity study of the inundation length and shoreline velocity (Fig. 5). It is the velocity component which is the governing variable of the required grid resolution. Since the 5,000 and 6,000 node solutions yielded similar shoreline motion behavior, the following results will be based on the 5,000 node FD LB solver. Figures 6(a–d) to 7(a–d) show the FD LB time dependent trajectories for the free-surface elevation and corresponding velocities. We observe the highly nonlinear wave transformation patterns from the (a) offshore to (d) shoreline motion location. For example, Figs. 7(a–d) illustrate further the highly nonlinear trajectories of the shoreline variables. The ones offshore exhibit non-circular trajectories which are transformed drastically to near disappearing ones, as the wave approaches the maximum runup location. Figures 8(a–c) show the FD LB solutions with the second-order NLSW and Boussinesq solutions produced by Pedersen (Chapter 17 in this volume). Pedersen models are based on a time step of 0.09 s and about 3171 nodes uniformly distributed. The three selected time instances represent maximum runup ($\approx$220 s), maximum run-down ($\approx$175 s) and near maximum run-down with maximum offshore velocity ($\approx$160 s). Key data from the models are outline in Table 1.

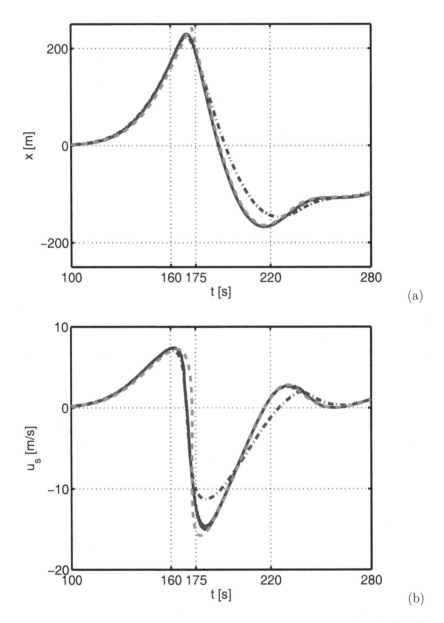

Fig. 4. Comparison of standard LBGK and FD LBM solutions. $-\cdot-$, Standard LBGK; $-$, FD LB solutions; $--$, Carrier et al. (2003).

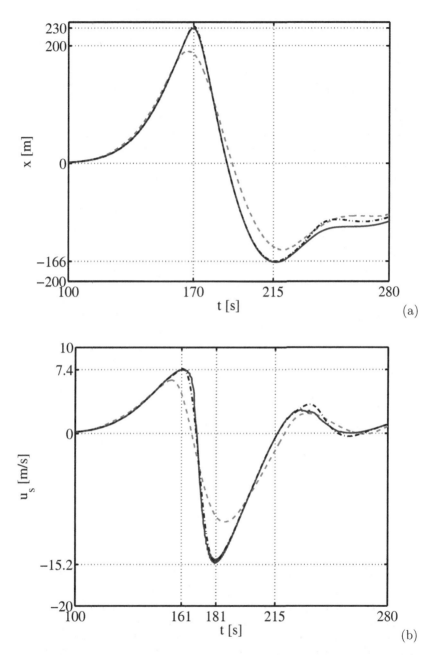

Fig. 5. FD LB solutions. Grids: – –, 1000 nodes; –, 5000 nodes; – · –, 6000 nodes.

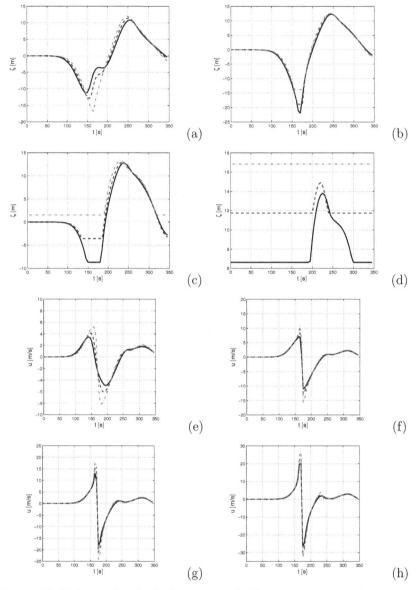

Fig. 6. FD LB free-surface (a–d) elevation and (e–h) velocity solutions of wave form transformation $x \in [-166, 500]$ m where $x = -166$ m and $x = +230$ m represent maximum runup and run-down locations. (a) Offshore, (b–c) Approaching shoreline, (d) The shoreline motion behavior towards maximum runup. (a,e) –, 500 m; – –, 400 m; – · –, 300 m. (b,f) –, 230 m; – –, 200 m; – · –, 150 m. (c,g) –, 100 m; – –, 50 m; – · –, 0 m. (d,h) –, −50 m; – –, −100 m; – · –, −166 m.

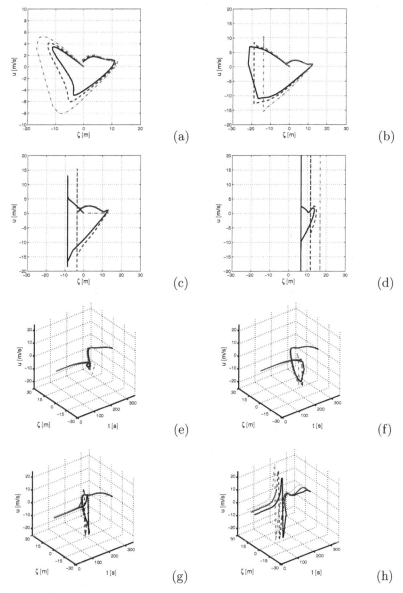

Fig. 7. FD LB velocity and free-surface trajectories of wave form transformation $x \in [-166, 500]$ m where $x = -166$ m and $x = +230$ m represent maximum runup and run-down locations. (a) Offshore, (b–c) Approaching shoreline, (d) The shoreline motion behavior towards maximum runup. (a,e) –, 500 m; – –, 400 m; – · –, 300 m. (b,f) –, 230 m; – –, 200 m; – · –, 150 m. (c,g) –, 100 m; – –, 50 m; – · –, 0 m. (d,h) –, −50 m; – –, −100 m; – · –, −166 m.

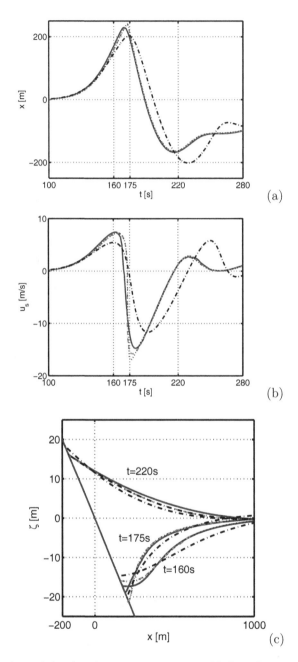

Fig. 8. Snapshots of shoreline location movement. −, FD LB solutions; − −, Carrier et al. (2003); · · ·, NLSW: Pedersen (Chapter 17); − · −, Boussinesq: Pedersen (Chapter 17).

Table 1. Model comparison of key data. LB solutions include shoreline algorithm of Lynett et al. (2002).

Maximum	Lattice Boltzmann Standard LBGK	Lattice Boltzmann 2nd-Order FD	NLSW (Pedersen (Chapter 17))	Semi-Analytical (Carrier et al. (2003))
Runup [m]	-147 ($t = 226$ s)	-166 ($t = 215$ s)	-164 ($t = 216$ s)	-164 ($t = 217$ s)
Run-down [m]	$+226$ ($t = 170$ s)	$+230$ ($t = 170$ s)	$+239$ ($t = 172$ s)	$+242$ ($t = 173$ s)
Shoreward velocity [m/s]	-11.29 ($x = 130$ m)	-15.22 (x,t) = (181 m,180 s)	-17 ($x = 176$ m)	-15.76 (x,t) = (174 m,177 s)
Offshore velocity [m/s]	$+7.09$ ($x = 179$ m)	$+7.44$ (x,t) = (161 m,162 s)	$+7.28$ ($x = 164$ m)	$+7.28$ (x,t) = (164 m,162 s)
No. of nodes	100,000	5,000	3,171	
Δx [m]	0.5	10	15.78	
Δt [s]	0.002	0.00015	0.09	
τ [–]	0.732	0.5	—	
CPU time [s]	6702	1975	$\approx$30	

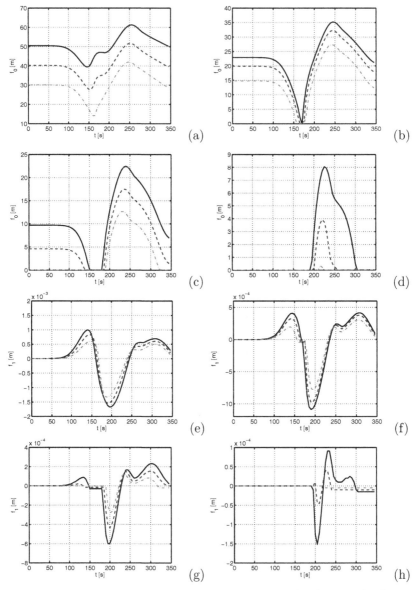

Fig. 9. The temporal variations of the distribution function (a–d) f_0 and (e–h) f_1 for $x \in [-166, 500]$ m where $x = -166$ m and $x = +230$ m represent maximum runup and run-down locations. (a,e) Offshore, (b–c, f–g) Approaching shoreline, (d,h) The shoreline motion behavior towards maximum runup. (a,e) –, 500 m; – –, 400 m; – · –, 300 m. (b,f) –, 230 m; – –, 200 m; – · –, 150 m. (c,g) –, 100 m; – –, 50 m; – · –, 0 m. (d,h) –, −50 m; – –, −100 m; – · –, −166 m.

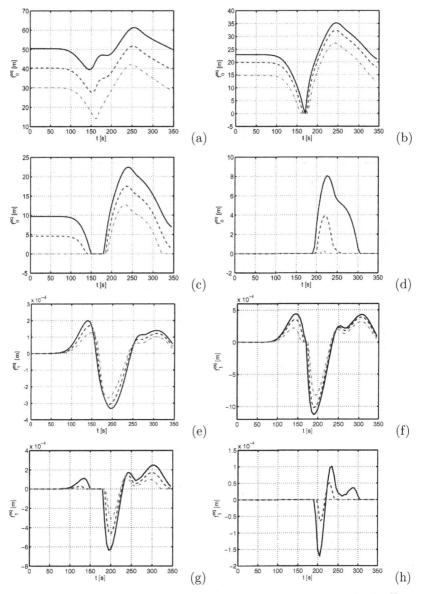

Fig. 10. The temporal variations of the equilibrium distribution function (a–d) f_0^{eq} and (e–h) f_1^{eq}. (a,e) Offshore, (b–c, f–g) Approaching shoreline, (d,h) The shoreline motion behavior towards maximum runup. (a,e) —, 500 m; – –, 400 m; — · —, 300 m. (b,f) —, 230 m; – –, 200 m; — · —, 150 m. (c,g) —, 100 m; – –, 50 m; — · —, 0 m. (d,h) —, −50 m; – –, −100 m; — · —, −166 m.

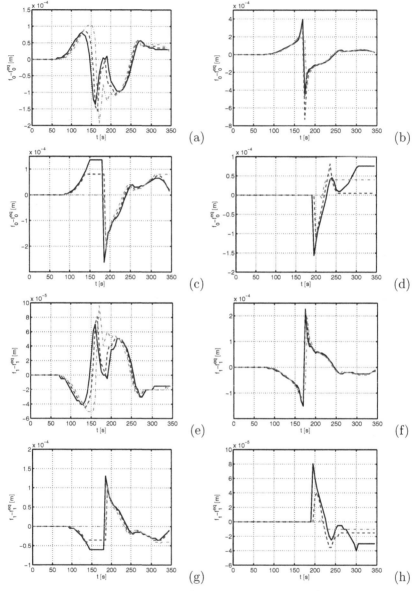

Fig. 11. The temporal variations of the collision term (a–d) $f_0 - f_0^{eq}$ and (e–h) $f_1 - f_1^{eq}$. (a,e) Offshore, (b–c, f–g) Approaching shoreline, (d,h) The shoreline motion behavior towards maximum runup. (a,e) —, 500 m; – –, 400 m; — · —, 300 m. (b,f) —, 230 m; – –, 200 m; — · —, 150 m. (c,g) —, 100 m; – –, 50 m; — · —, 0 m. (d,h) —, −50 m; – –, −100 m; — · —, −166 m.

The FD LB and NLSW solution of Pedersen (Chapter 17) compares well with the semi-analytical solution of Carrier et al. (2003) (as expected, since they are all NLSW models) whereas the Boussinesq model seems more inaccurate for the same grid refinement. Any discrepancy and difficulty for all models were mainly related to the large accelerations at maximum run-down. We looked further into the variation of the collision term, as this would be a means of examining free-surface and shoreline nonlinearity. Figures 9(a–d) to 11(a–d) show the time series of f_i, f_i^{eq} and $f_i - f_i^{eq}$ for the discrete microscopic velocities c_0 and c_1. We can observe that f_0 and f_0^{eq} follow similar patterns as the free-surface elevation ζ whereas f_1 and f_1^{eq} tend to follow the macroscopic behavior of the velocity u. The

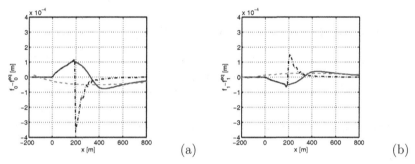

Fig. 12. Snapshots of collision term variations for the discrete particle velocities (a) c_0 and (b) c_1. —, $t = 160$ s; — · —, $t = 175$ s; — —, $t = 220$ s.

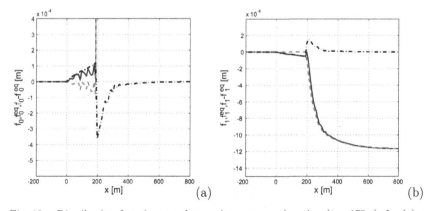

Fig. 13. Distribution functions at the maximum runup location ($t = 175$ s), for (a) c_0 and (b) c_1. —, f_i; — —, f_i^{eq}; — · —, $(f_i - f_i^{eq})$.

collision term $f_i - f_i^{eq}$ does not follow any macroscopic patterns. Due to the coarse grid, we can unfortunately not make any further interpretation in terms of bridging the behavior of the mesoscopic and macroscopic variables. Figure 12 shows the variations of the collision term $(f_i - f_i^{eq})$ for the discrete microscopic velocities c_i at $t = 160$ s, 175 s, and 220 s. We can observe that the deviations from equilibrium are largest in the run-down region near the occurrence at maximum velocity. This is further confirmed in curves of Fig. 13 and thus agrees with the observations of the macroscopic variables. Finally, we should note that the LB maximum velocity, maximum runup and run-down occurred at slightly different instances than the semi-analytical solution of Carrier et al. and thus the LB extreme values would deviate from the solutions shown.

9. Other Beaches and Initial Wave Forms

We extended the benchmark test 1 to study the wave runup/run-down effects when varying the (1) beach slope, and (2) initial wave form. The steepest slope case (1/5) shows the smallest runup/run-down where the opposite was the case for the flattest slope (1/20), as expected (Fig. 14). We were somewhat surprised by the similar form of the inundation length

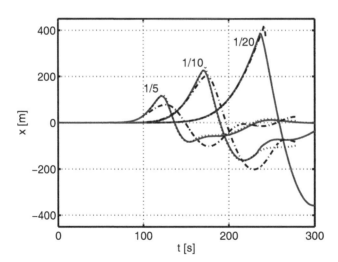

Fig. 14. Slope sensitivity. —, FD LB solutions; ···, NLSW: Pedersen (2007); – · –, Boussinesq: Pedersen (2007).

for all three slopes of the FD LB model. For example, one would expect that the long wave models would be less accurate for relatively steep slopes, however, the steepest slope of 1/5 we tested showed good agreement between the LB and the NLSW solutions. The largest discrepancy occurred at maximum run-down. The grid convergence study of Pedersen (Chapter 17) showed no runup, i.e. breaking around maximum withdrawal where predicted with minor differences between the NLSW and Boussinesq solutions when $\partial h_b/\partial x = 1/20$ whereas the LB model did predict shoreline motion profile similar to the 1/10 case (just enlarged in shape). It is unclear why

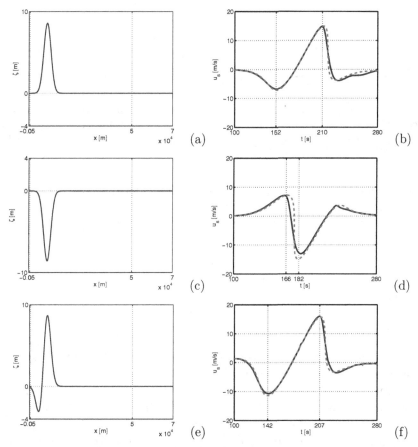

Fig. 15. Shoreline velocity time history (b,d,f) based on initial wave forms (a,c,e). —, FD LB solutions; – –, Carrier et al. (2003).

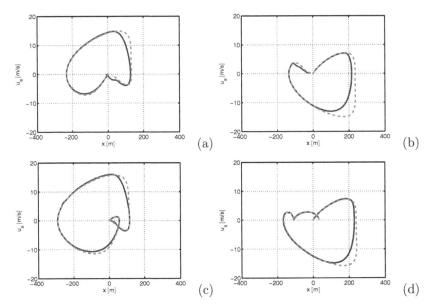

Fig. 16. Shoreline velocity versus shoreline location. (a) Initial Gaussian wave form, case a, (b) negative Gaussian shape, case b, (c) leading depression N-wave, case c, (d) waveform caused by submarine landslide, case d (bench mark test 1). –, FD LB solutions; – –, Carrier et al. (2003).

Pedersen's model solutions did not predict runup for the slope of 1/20. The other beach slope cases showed similar trajectory effects on the shoreline motion when comparing the LB with the NLSW and Boussinesq solutions (Pedersen (Chapter 17)).

Finally, we tested the LB solver for other initial wave forms, after Carrier et al. (2003). We found fairly good agreement for the 5000 LB node solutions, as shown in Figs. 15 and 16.

10. Concluding Remarks and Future LBM Endevours

We have tested two LB schemes which include a standard LBE formulation and a second-order accurate FD LB model. We investigated two approaches to treat the wet-dry interface: (1) a thin film layer on the beach, and (2) the shoreline algorithm of Lynett et al. (2002). Our model is a 1-D idealization. The FD LB scheme with the algorithm of Lynett et al. agreed well with results of other investigators. This includes all four case studies, as described by Carrier et al. (2003). We found that the FD LB solver (5000 nodes) is competitive for tsunami predictions, as the CPU is about 30 min on a single processor 3 GHz on a standard PC of today. The CPU of the LB model is, however, not as fast as the NLSW/Boussinesq models presented by Pedersen (Chapter 17). The LB model required higher resolution and the additional velocity set in the luggage. The standard LBE solver with the shoreline algorithm of Lynett et al. (2002) also performed reasonably well. But is was less accurate and less efficient than the FD LB solver. About 100,000 nodes was required and the CPU was about 1 hour 50 min. None of the LB schemes produced accurate velocity without a shoreline algorithm, i.e. the thin film approach did not work well. We should note, however, that the inundation length was agreeing well with other investigators' solutions.

The present Boltzmann modeling approach may be a new competitive candidate for predictions of free-surface water waves and nearshore processes, especially where high density resolution would be required. Obviously unstructured lattices are highly desirable for the runup/run-down problems and could be a simple explanation for not achieving better agreement for the LB schemes without a shoreline algorithm. From "a big picture" perspective, the LB solutions herein may be viewed as a special finite-difference discretization of the depth-averaged transport equations. The LB solutions herein do not offer anything new on the details of physical processes during runup/run-down on the beach, compared to traditional NLSW models, due to the coarse and simple 1-D model. Like any other numerical model the LB solutions should most likely be obtained on unstructured fine resolution lattices in the near shore region. Reaching this level of the LB development, we would be in a better position to conclude whether or not this model approach would offer further insight into the physics more than other current methods. But at least for now, one has hopefully convinced the reader that a method with roots in gas dynamics could be applied to water wave problems of this kind. Regarding future research, we hope that it can work for breaking waves; the original idea/goal of using the LB approach.

There are several other fundamental issues which need further investigation. First, the approximation of the collision operator should be improved through a Multi-Relaxation-Time (MRT) scheme in which the relaxation times for different kinetic modes are separated (τ is no longer constant). This would offer realistic modes between particles and perhaps reveal the deeper physics we are looking for. Second, grid efficiency in terms of irregular denser resolution in the nearshore region could be achieved using multi-block algorithm similar to that of Yu et al. (2002). Furthermore, there exists several unresolved issues in the solutions presented both at the near shore and at the open boundary. The near shore physics could be represented better, as the bed friction and sediment transport effects are not included in our results. Nor is the effect of wave breaking considered. To delay wave reflections at the open boundary, we adopted a simple extrapolation equation at the open boundary and extended the flow domain with 500 m to mimic some form of sponge layer. Other algorithms could be explored to improve accuracy and efficiency.

In a final remark, we should emphasize that our intention with the LB modeling approach is to use it in local areas where physical details are needed such as the nearshore regions. Therefore a realistic goal and potential improvement to the current tsunami model literature would be to propose a LB-model coupled with a macro-scale level model for large domain tsunami model predictions.

We attempted to write-up the LB solutions of benchmark test case 1 as self-contained as possible. We realize, however, the shortcomings of explanations, and would like to refer the reader to further discussions, details and references about the LB method which are given in Chapter 5 in this volume.

References

Balzano, A. (1998) Evaluation of methods for numerical simulation of wetting and drying in shallow water flow, *Coastal Engineering* **34**, pp. 83–107.

Bhatnagar, P., Gross, E. and Krook, M. A. (1954) A model for collision processes in gases i: small amplitude process in charged and neutral one component systems, *Physics Review A* **94**, pp. 511–526.

Carrier, G. F., Wu, T. T. and Yeh, H. (2003) Tsunami run-up and draw-down on a plane beach, *Journal of Fluid Mechanics* **475**, pp. 79–99.

Frandsen, J. B. (2008) "Free-surface Lattice Boltzmann modeling in single phase flows", in *Advanced Numerical Models for Simulating Tsunami Waves and Runup*, Advances in Coastal and Ocean Engineering, Vol. 10, eds. Liu, P. L.-F., Yeh, H. and Synolakis, C. (World Scientific Publishing Co.).

Gombosi, T. I. (1994) *Gaskinetic Theory* (Cambridge University Press).

Guo, Z. and Zhao, T. S. (2003) Explicit finite-difference lattice Boltzmann method for curvilinear coordinates, *Physical Review E* **67**, 066709.

Junk, M., Klar, A. and Luo, L.-S. (2005) Asymptotic analysis of the Lattice Boltzmann equation, *Journal of Computational Physics* **210**, pp. 676–704.

Kennedy, A. B., Chen, Q., Kirby, J. T. and Dalrymple, R. A. (2000) Boussinesq modeling of wave transformation. breaking and runup. I:1D, *Journal of Waterway, Port, Coastal, and Ocean Engineering* **126**(1), pp. 39–47.

Lynett, P. J., Wu, T.-R. and Liu, P. L.-F. (2002) Modeling wave runup with depth-integrated equations, *Coastal Engineering* **46**, pp. 89–107.

Mei, R. and Shyy, W. (1998) On the finite-difference-based Lattice Boltzmann method in curvilinear coordinates, *Journal of Computational Physics* **143**, pp. 426–448.

Pedersen, G. (2008) "A Lagrangian model applied to runup problems", in *Advanced Numerical Models for Simulating Tsunami Waves and Runup*, Advances in Coastal and Ocean Engineering, Vol. 10, eds. Liu, P. L.-F., Yeh, H. and Synolakis, C. (World Scientific Publishing Co.).

Prasad, R. S. and Svendsen, I. A. (2003) Moving shoreline boundary condition for nearshore models, *Coastal Engineering* **49**, pp. 239–261.

Succi, S. (2001) *The Lattice Boltzmann Equation for Fluid Dynamics and Beyond* (Oxford University Press).

Yu, D., Mei, R. and Shyy, W. (2002) A multi-block Lattice Boltzmann method for viscous fluid flows, *International Journal for Numerical Methods in Fluids* **39**, pp. 99–120.

Yu, D., Mei, R. and Shyy, W. (2005) Improved treatment of the open boundary in the method of the Lattice Boltzmann equation, *Progress in Computational Fluid Dynamics* **5**(1/2), pp. 3–12.

CHAPTER 17

A LAGRANGIAN MODEL APPLIED TO RUNUP PROBLEMS

G. Pedersen

Mechanics Division, Department of Mathematics, University of Oslo
E-mail: geirkp@math.uio.no

1. The Models

The main runup model of this section is based on Lagrangian Boussinesq equations that are fully nonlinear and possess standard dispersion properties. It is run in both hydrostatic (NLSW) and dispersive mode. For comparison we employ also an Eulerian FDM code for the standard Boussinesq equations and a Boundary Integral Method (BIM) for full potential theory. The first is used for linear computations and is without any particular runup feature; runup heights are found from the vertical elevation at the shore. The BIM is based on Cauchy's formula, Lagrangian surface nodes and spline interpolation. All models allow for a variable spatial resolution, but this feature is employed only for the BIM and the Eulerian models. Systematic grid refinement is employed for all applications reported.

Expressed in terms of the depth averaged velocity, $\bar{u}$, and the total depth, H the Lagrangian long wave equations read

$$H\frac{\partial x}{\partial a}=H_0 \text{ (a)}, \quad (1-r_1)\frac{\partial \bar{u}}{\partial t}=-g\frac{H}{H_0}\frac{\partial H}{\partial a}+g\frac{\partial h}{\partial x}-r_2+S_1+S_2 \text{ (b)},$$

$$S_1=\frac{1}{2}H\frac{\partial^3 h}{\partial t^2 \partial x}, \quad S_2=-\frac{1}{g}\left(2\bar{u}\frac{\partial^2 h}{\partial t \partial x}+\frac{\partial^2 h}{\partial t^2}\right)\frac{\partial \bar{u}}{\partial t}-\frac{\partial^2 h}{\partial t \partial x}\frac{\partial H}{\partial t}+H\bar{u}\frac{\partial^3 h}{\partial t \partial x^2},$$

(1)

where a is the Lagrangian coordinate, h is the equilibrium depth and $H_0 = H(a,0)$. The position x relates to the velocity according to $\partial x/\partial t = \bar{u}$. Only the shallow water (hydrostatic) terms are spelled out in (1b). A more complete description of the equations in absence of a slide, including the dispersion terms (r_1 and r_2), is found in Jensen et al. (2003)[2] and references therein. The principal forcing due to a time dependent bottom is the modified source of horizontal momentum from the second term on the right hand side of (1b). Higher order source terms, S_1 and S_2, have been included particularly for benchmark 3. Replacing H by h in S_1 we obtain the linear part that is employed in the Eulerian Boussinesq equation, while S_2 is exclusively nonlinear.

A moving shoreline is associated with a fixed a, where the condition $H = 0$ is employed. The set (1) is solved by finite differences on a grid that is staggered in space and time. The hydrostatic (NLSW) version is explicit, while dispersion requires implicitness. No smoothing or filtering is employed with the long wave models.

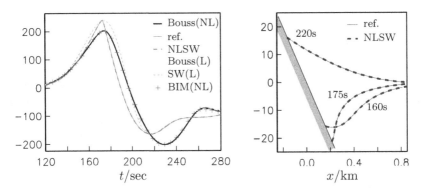

Fig. 1. Left: withdrawal (m). Right: surface (m) at selected times.

2. Benchmark Problem 1

The reference solution[1] exists in a semi-infinite domain, but have been made available until 50 km from the coast. We add a deep water region ($x > 50$ km) to avoid that reflections from the seaward boundary reach the shore region for $t < 280$ s. The alternatives are open boundary conditions or sponge layers that may introduce additional errors and uncertainties.

The inundation lengths shown in Fig. 1 are all close to convergence (increment $\Delta a = 16$ m for the nonlinear long wave models). The NLSW solution agrees very well with the reference solution (ref.). Dispersion has some effect and reduces the draw-down and increases the runup. The maximum runup and drawdown are nearly equal for linear and nonlinear models.

Since the initial condition inherits very small elevation/depth ratios this is to be expected in the hydrostatic approximation[4]. That also the linear (L) and nonlinear (NL) Boussinesq models yield nearly identical extrema indicates that dispersion and nonlinearity are important in the deep and shallow regions, respectively, with no or little overlap.

The Boussinesq and full potential (BIM) models yield nearly identical solutions near shore, while there are small discrepancies in the waves propagating into deeper water (not shown).

From a modeling point of view the NLSW solution is the most interesting (challenging) one. At maximum drawdown there is a (near) cusp on the inundation curve corresponding to huge accelerations, a solution that is close to being "weak" and slow convergence of the numerical model. In Fig. 2 we observe that the accuracy for the velocity is much poorer than for the inundation length. The convergence of the numerical solution is particularly slow close to $t = 173$ s where the reference solution is slightly multi-valued. These features have not been pursued further. Moreover, for the maximum withdrawal the NLSW model displays a linear convergence except for extremely fine resolutions. For the smooth Boussinesq solution we obtain quadratic convergence (dashes in right panel).

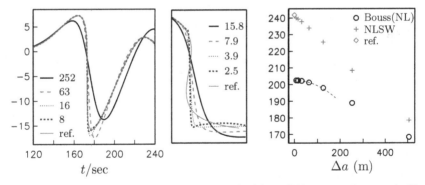

Fig. 2. Left and mid panels: shoreline velocities (m/s). NLSW computations, marked by Δa (m), are compared to the reference solution. Mid panel: blow-up of region $(172, 176) \times (-18, 5)$. Right: Convergence of maximum withdrawal (m).

3. Benchmark Problem 3

In this problem[3] a slide with time dependent shape and acceleration penetrates the water. All results for this benchmark is given in meters and dimensionless time units, $\sqrt{\delta/g\mu^2}$, as defined in the reference. There is a singularity at the rear of the slide body ($x = 0$), while it is undefined for $x < 0$. Hence, if the fluid reaches $x = 0$ before the height of the slide body becomes negligible at this point, conceptual and practical problems arise in the numerical solution. For case A this occur before the first runup maximum. This is not so for case B for which the NLSW and Boussinesq(*) equations yield:

Δa	0.16*	0.16	0.31	0.63	2.51	5.04	10.17
Runup	0.495	0.495	0.495	0.496	0.491	0.458	0.468
Drawdown	−0.235	−0.233	−0.231	−0.226	−0.207	−0.221	−0.318

The solutions for the different models are shown in Fig. 3. For case A and $t = 1.5$ the reference solution agree closely with the nonlinear solutions except for a small discrepancy very close to the shore. The differences between the linear and nonlinear solutions are significant for case B at $t = 2.5$, while effects of dispersion are visible only at the front of the offshore propagating wave. The Eulerian model SW(L) is close to the reference solution for both cases (difference up to 0.0025 m with shoreline resolution $\Delta x = 15$ m for case A and $t = 1.5$). For case B inclusion of the dispersive source term, S_1,

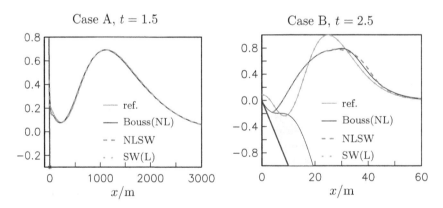

Fig. 3. η (m); comparison of models.

reduces the maximum height of the outgoing wave from 1.42 m to 1.33 m at $t = 4.5$. S_2 then increases the height by 0.01 m. For case A and $t = 1.5$ the term S_1 reduces the wave height from 0.698 m to 0.692 m, while the effect of S_2 is negligible (less than 10^{-5} m).

4. Remarks

Lagrangian (and ALE) models are undoubtedly useful for idealized studies, as benchmarks 1 and 3, where they produce very accurate results. However, such models have poorer prospects for complex tsunami studies. Further discussion and references on related models are given in Chapter 1 in this volume (Pedersen)[4].

References

1. G. F. Carrier, T. T. Wu and H. Yeh, *J. Fluid Mech.* **475**, 79 (2003).
2. A. Jensen, G. Pedersen and D. J. Wood, *J. Fluid. Mech.* **486**, 161 (2003).
3. P. L.-F. Liu, P. Lynett and C. E. Synolakis, *J. Fluid Mech.* **478**, 101 (2003).
4. G. Pedersen, Modeling runup with depth integrated equation models, in *Advanced Numerical Models for Simulating Tsunami Waves and Runup*, eds. P. L.-F. Liu, H. Yeh and C. Synolakis (World Scientific Publishing Co., 2008), p. 3–41.

Appendix

The following article was omitted from Volume 9 of the review series, entitled "PIV and Water Waves" (eds. J. Grue, P. L.-F. Liu and G. K. Pedersen).

APPENDIX

PHASE-AVERAGED TOWED PIV MEASUREMENTS FOR REGULAR HEAD WAVES IN A MODEL SHIP TOWING TANK

Joe Longo, Jun Shao, Marty Irvine, Lichuan Gui and Fred Stern

IIHR Hydroscience and Engineering
300 S. Riverside Drive
The University of Iowa, Iowa City, IA, USA 52242
E-mail: joseph-longo@uiowa.edu

Phase-averaged towed PIV measurements are made for regular head waves in a model ship towing tank using both fixed and towed PIV systems in preparation for tests with a model ship. Data reduction and uncertainty assessment software are developed and tested for calm water and long-wave, low steepness conditions used in previous forces, moment, and wave pattern tests with surface combatant. The results are validated through comparisons with progressive wave theory.

1. Introduction

Focus of experimental ship hydrodynamics research is moving into unsteady viscous flows in support of unsteady Reynolds-averaged Navier-Stokes (RANS) code development for simulation-based design. The forward-speed diffraction problem[1], i.e. restrained body advancing in regular headwaves, is a building block problem towards ultimate goal of physical understanding and simulation of viscous nonlinear seakeeping and 6DOF maneuvering. The present study is precursor for measurement of the unsteady nominal wake of a naval combatant and provides documentation of the data-acquisition and reduction procedures along with detailed data of the incident headwave and comparisons with 2D progressive-wave theory.

2. Test Design

The tests are conducted in the 100 × 3 × 3 m IIHR tank, which is equipped with a drive carriage and trailer, wavemaker, and moveable wave dampeners.

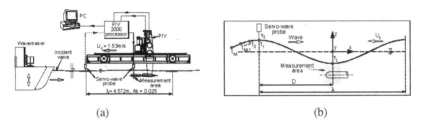

(a)   (b)

Fig. 1. Experimental setup for towed, unsteady PIV measurements.

The DANTEC, towed, 2D PIV measurement system[2] is linked to a servo wave gage positioned either directly over or upstream of the measurement area to phase lock the PIV measurements (Fig. 1(a)). The system is configurable to measure in vertical (xz; Fig. 1(b)) and horizontal (xy) planes. A separate measurement system is used for carriage speed. All tests are performed with $U_c = 1.53$ m/s and without $U_c = 0$ m/s forward speed for cases with and without waves. Headwave parameters are wavelength $\lambda = 4.572$ m, frequency $f_w = 0.584$ Hz, and steepness $Ak = 0.025$ and frequency of encounter $f_e = 0.922$ Hz, which are all based on previous studies[2,3,4]. Measurement area dimensions are 192 × 1018 pixels or 14.3 × 74.9 mm. PIV image pairs are taken at 133 ms intervals with $\Delta t = 490$ μs between images. Data is acquired at $z = -25.0, -53.34, -110.45$ mm and $z = -25.0$ mm for xz and xy configurations, respectively. Datasets are acquired at each elevation and configuration through repeated carriage runs, stockpiling 1200 and 2000 vector maps for steady and unsteady cases, respectively, to obtain convergence. Data-reduction software was developed to perform harmonic data analysis and includes procedures for phase-sorting a dataset, two-stage filtering for spurious vectors, 5th order least-squares curvefitting to filtered data, and 2nd order Fourier series analysis of least squares. Uncertainty assessment (UA) follows standard procedures[5].

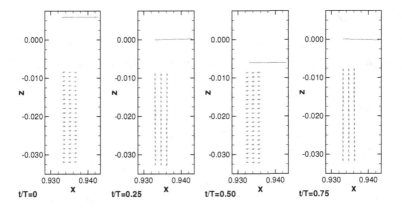

Fig. 2. Vector field for four time instances in one encounter period.

3. Results and Comparison Theory

Figure 2 illustrates experimental wave elevation and vector fields for $z = -25.0$ mm at four instances in one encounter period (T). Expected wave-induced velocities are displayed and correlate with free surface elevation. Quantitative validation is performed through comparisons of harmonic variables of 2D progressive-wave theory and experiment. For $z = -25.0$ mm, variables are averaged through the measurement area and experiment is subtracted from theory and expressed as a percentage. Average level of agreement for 0th and 1st harmonic amplitude is 0.3% and 0.8% for U, respectively, and 0.1% and 0.8% for W, respectively. For 1st harmonic phase, agreement is 1–2% for both U and W which is roughly 3–7°. Average turbulence is 0.01% (i.e. 0th harmonic uu_0, ww_0, uw_0), however, 1st harmonic amplitudes of turbulence are small but significant (0.005%), near the location of the incident wave trough.

4. Future Work

Future work consists of steady and unsteady PIV measurements and UA of the nominal wake plane of model 5512[6] for the same conditions presented herein. The data and UA will be archived at www.iihr.uiowa.edu/~towtank.

Acknowledgments

This research was sponsored by the Office of Naval Research under Grant N00014-01-1-0073 under the administration of Dr. Pat Purtell.

References

1. H. Rhee and F. Stern, *Int. J. Num. Meth. Fluids*, **37** (2001) 445.
2. L. Gui, J. Longo and F. Stern, *Exp. Fluids*, **31** (2001) 336.
3. L. Gui, J. Longo, B. Metcalf, J. Shao and F. Stern, *Exp. Fluids*, **31** (2001) 674.
4. L. Gui, J. Longo, B. Metcalf, J. Shao and F. Stern, *Exp. Fluids*, **32** (2002) 27.
5. Test uncertainty: Instruments and apparatus, *ASME PTC (Performance Test Code)* 19.1–1998 (1998), 112.
6. J. Longo, J. Shao, M. Irvine and F. Stern, *24th Symposium on Naval Hydrodynamics* (2002).